PEOPLE FIRST

Unite your tech teams to deliver successful M&A integration projects

HUTTON HENRY

First published in Great Britain 2018
by Rethink Press (www.rethinkpress.com)

CONTENTS

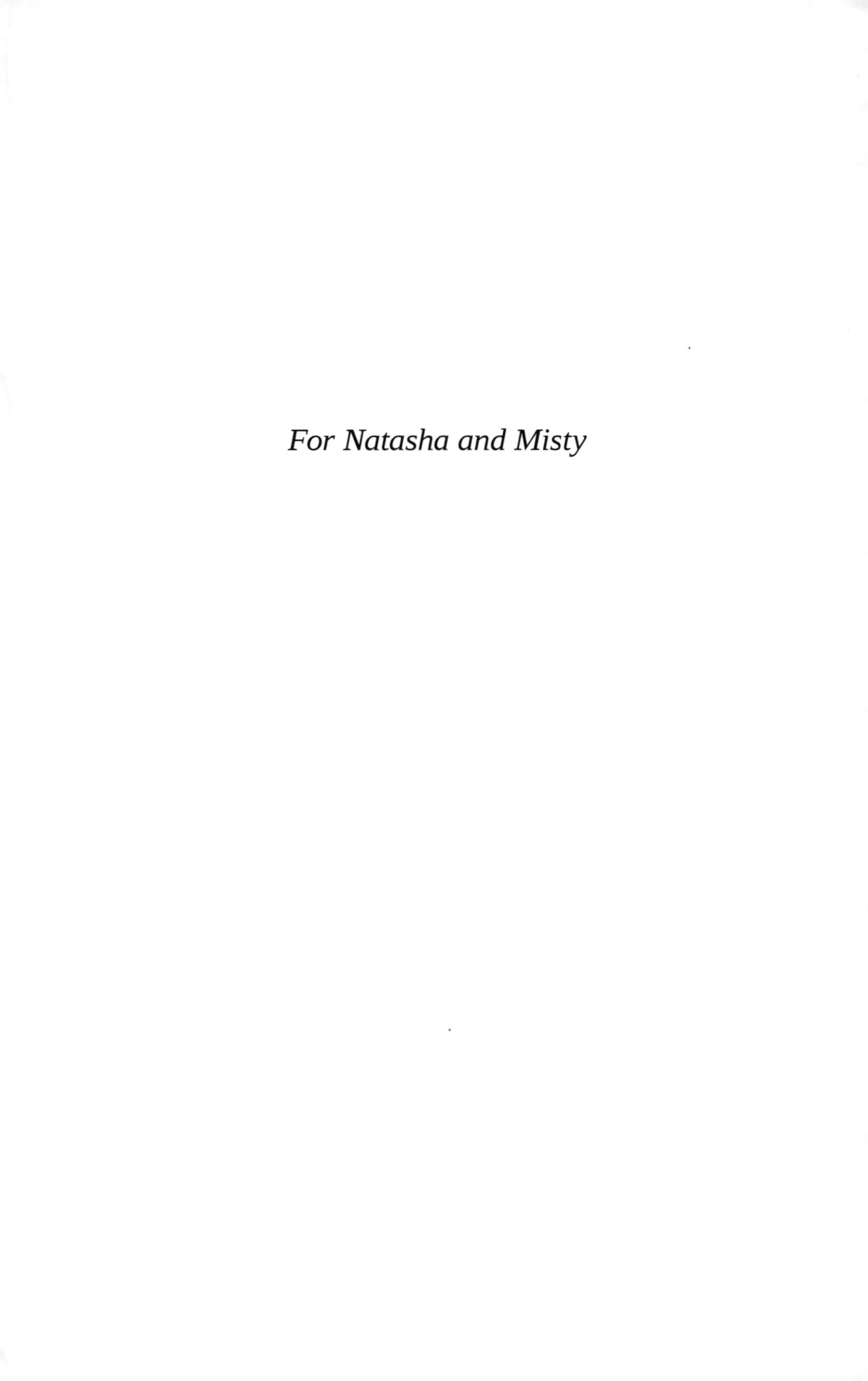

For Natasha and Misty

Introduction

It has become appallingly obvious that our technology has exceeded our humanity.

ALBERT EINSTEIN

Why read this book?

Laos, South-east Asia, a long time before the internet landed on the backpacker trail. My friends and I had just spent the day 'tubing': travelling on enormous tractor tyres across the Mekong River. It had been exhilarating when it was fast, and tranquil when both the river and the sun decided to take it easy.

After tubing finished, we did the tourist thing of exploring pitch-dark caves up to our waists in water. There were about twenty of us, young and eager, and the tour guide fancied my close friend. Maybe that's why he gave me the candle, to get closer to her, I certainly didn't want it. If I am honest, my excitement turned to concern as I had to walk through damp, dark caves, providing light and direction for the entire group.

I was responsible.

If you have a post-merger integration (PMI) project on the horizon, you've got that candle coming. You already have a significant responsibility on your shoulders.

If you represent the business making the acquisition, the project will have a major impact on the company you work for, its growth plans, the futures of the employees in all the companies involved, and, last but not least, your career progression. If you work for a consultancy, lucky you. Not only do you have to deliver to your client, but you also have the challenge of getting to know both your client and the business to be acquired while ensuring the project meets the goals of the client and the firm you work for. Either way, you and your integration team will have a significant impact on people.

By implementing the steps outlined in this book, you will be able to reignite the enthusiasm in disengaged technology employees, arrange them into more effective teams, then plan and orchestrate your project – all of which will benefit your career and future opportunities. *People First* focuses on the early stages of a PMI project: the period that will lay the foundations for the future success of a lot of people. But you can't run a successful, in-depth 'people transformation' programme unless you have applied the same process to yourself.

I also believe that reducing staff attrition and negativity, which often haunts these projects, is an essential key deliverable of these programmes. Often, my 'people focus' is seen as an unnecessary concern. 'We can replace anyone who leaves,' I'm told, but it is key to remember that:

a. loss of key staff or knowledge can impact the effectiveness or growth, therefore the value, of an acquisition
b. as explained in the Engage section, it costs at least 125% of an employee's salary to replace someone who leaves a business

So it makes sense to start with prevention of these issues in mind.

Because the subject is a little dry, I've woven in travel stories to illustrate my points. I chose travel stories because they involve situations that are unusual for all of us, where routines are broken. Most of my stories involve other people – often strange people.

Let's go back to the Laos situation. I was wearing flip-flops, I was submerged in water up to my waist, and of course, I slipped and lost my grip. 'Sssst' – the candle went out and we were in pitch darkness. I guess I should have been worrying about the team, but at the time I could only focus on my own embarrassment. But we pressed on and got there in the end. And that's the reality of projects, regardless of the situation – with others in the right positions on board, you'll always get there.

Who this book is for

This book is for leaders within corporate technology teams who are embarking on a project related to a corporate merger or divestment.

I've found that similar people issues occur within high-growth tech start-ups. While *People First* is primarily aimed at leaders in technology, it will benefit anyone interested in team dynamics within a technology team, including leaders within Human Resources, Learning and Development and business Strategy/ Operational teams.

To keep the content focused on the people + technology aspect, I have purposely not used industry or academic case studies in this book. There are plenty of books that provide solid case studies, such as *Why Deals Fail* by Faelten, Driessen and Moeller, or *M&A IT Best Practices* by Janice M. Roehl-Anderson, and the industry constantly publishes case studies online. Another source is Declan Burke's excellent 'IT Integration in Mergers and Acquisitions' available here: http://ccsenet.org/journal/index.php/ijbm/article/view/70146/38895

My story

When I was five, I lost my parents, was separated from my siblings, and was placed with numerous foster care families until I finally settled. I still recall my first day in what became my home for thirteen years, sitting at the dinner table in a strange new world. I was nervous, lonely and scared; I wasn't sure what I should be doing, so I did what I've continued to do to this day. I got stuck in. As everyone finished their dinner, I pulled a chair over to the sink and stood on it to help with the washing-up.

Sadly, my foster parents did not encourage education. Luckily, though, this was the 1970s. I discovered a passion for computers that allowed me to self-teach and escape into a world of creativity. My (blood) brother, Trevor, on leave from the Air Force, gave me my first computer when I was eight, and the rest is history. I devoured every bit of information I could find and spent my time trying to code.

I never studied at school and found ways of avoiding homework throughout my entire education, but I had a great time with friends. I also started my first 'business' at school – the more affluent kids would discard their designer school blazers while they still had some wear in them, so I would resell them to the less fortunate. You have to make the best of a bad situation.

While education in my foster home was lacking, I greatly appreciate the care I was given and the roof I had over my head. However, I left school at sixteen with no qualifications, little confidence, and most of my teeth badly decayed. The anxiety this caused made it difficult for me to talk to others, let alone get a job.

Crippled by constant toothache, I went to see a dentist. He repaired my teeth and, in turn, my confidence. I was accepted on to a school leavers' scheme, and when I was eighteen, the boss broke the 'graduates only' rules and took a chance on me, giving me a graduate job.

Being cheeky, I said, 'I'll think about it,' which resulted in a £2,000 increase in my annual salary. I wish I had those balls now.

From there, my career and life evolved. I've been blessed. My brother, my dentist and my first boss all helped to change my life. Consequently, I've made an effort to help others whenever possible throughout my career.

So how does all this relate to M&A (mergers and acquisitions) and, more specifically, PMI? I have worked with numerous technology teams that have recently discovered their employer has been acquired by another business, and I can understand how they feel. It takes me right back to when I was five on my first day in care. Those employees' feelings of disorientation and lack of safety are all too familiar to me. In my opinion, though, it is best to see the transformation as an opportunity, not a threat. I work hard to help people do exactly that.

How to use this book

You can either read *People First* from cover to cover, or just dip in as needed. But I recommend you understand the 'Engage' phase as this underpins everything else.

In Part One, we will look at what an M&A PMI is and outline the challenges and risks, discussing how the world of IT is changing and examine the need for IT staff to adapt – quickly. Part Two will progress to the Managed M&A: an approach to delivering complex merger, acquisition or divestment projects to ensure that you:

- Create and retain a high-performing joint technology team that works more effectively and drives towards a highly successful outcome for all
- Reduce the significant risk of reduced profits and reputation damage caused by loss of key talent and valuable propriety knowledge
- Increase certainty from an organizational, team and individual perspective

The Managed M&A consists of three key phases:

Engage - where you develop the foundations to create a high-performing PMI team.

Examine - where you carry out IT due diligence with a 'people first' approach.

Envision - where you deliver solid project management and technical design plans specific to the type of transformation being undertaken.

A cautionary tale

It was a grey, rainy afternoon in the Midlands, UK in the mid-1990s. My team and I had found sanctuary in a taxi which was transporting us to a hotel. There was a lot of energy in the car; in fact, we'd not stopped speaking since we left London. We would be attending a meeting the following morning with some important people and we were well prepared. We were excited.

While the others were speaking and joking around, I caught the cab driver's eye in the rear view mirror. He was literally scowling at us. I couldn't understand what his concern was, so I shrugged it off and continued chatting with the team.

'Up from London?' he asked.

Someone politely confirmed that yes, we were.

The driver then launched into a barrage of abuse – about us; about 'Londoners' in general. The angrier he got, the more he sped up. Our attention shifted immediately from business to the driver's reckless behaviour.

Eventually, his rant arrived at his real concern: the corporate company we represented was evil. We were impacting him, his family, and everyone else in the area. At this point, we were locked in the taxi on a hair-raising journey as the car sped through the rain. Our requests for the taxi driver to slow down went unheard, and he continued to complain bitterly.

That was my first introduction to post-merger integration (PMI).

The company we represented had made a high-profile acquisition of another large corporate, and the local community was worried about the impact on their livelihoods. That week, we learnt that the community resented the buy-out, and were not scared to show it.

For the employees of the acquired company and the people who worked for their suppliers, major changes were ahead. People were worried. Even now, I remember arriving at the office, which was beige and lacking in any energy. It was soulless. The teams did not respond to our requests for help – we were unwelcome outsiders. This was my first global software project, implementing a new system that deployed leading-edge technology in order to mechanise and standardise, moving away from legacy mainframe solutions. But while I was busy learning how to manage projects, I was oblivious to the real lesson that lay ahead. The angry, frustrated taxi driver was just the beginning.

I had been working directly for the 'buy-side' business – the company that had made the acquisition – for over six years, so I knew the company that I worked for and the key people within it well. We had agreed to launch the new system in a newly acquired corporation because this would demonstrate the cost-saving benefits of a shared system. Even now I cringe when I think about this decision – a brand new, high-risk, high-impact system being rolled out in a newly acquired business.

Soon after the pilot site went operational, I was awoken by an IT director screaming down the phone, 'What the **** have you done?' Information from our software was being scrambled – not good for legal financial documents. We couldn't wait for a third party to develop a solution, so the team developed a temporary fix, but the post mortem was gruesome. I remember crying as I drove home. I took my job seriously.

When my team and I visited the users in the acquired business, we realised that all the people who'd been involved in our 'extensive' testing hadn't actually tested anything. The apathy in the team, caused by the acquisition, was so significant that they had signed off the system without checking the output. They simply didn't care.

Looking back, I now understand that they were blinded by fear. Fear of the change, the unknown, and the potential impact on their lives. Neither our team as employees of the buy-side company nor the people working for the acquired business had been well briefed about the acquisition. None of us had been helped to manage a multi-party, multi-technology project. We just knew we had to get on with the job in hand; we didn't know anything about the strategic fit, or the benefits of the M&A deal for either company, the customers, or us as individuals. To make matters worse, we had been working long hours, often staying away from home in order to deliver the IT programme for the business.

Having worked on various similar projects since then, I now realise that a short on-boarding exercise with parties from both companies will result in better collaborative working and better overall results. From the target side, the technology employees need to know what's in it for them. The buy-side company's employees also need to know what the all-round benefits will be because it is not human nature to work on projects that may create distress or issues for anyone.

To me, my first experience demonstrated the power of (a) being exposed to major projects and responsibility, and (b) allowing people, especially young people, to learn from their mistakes. Even now, I am grateful for opportunities to learn at work. Unfortunately, though, I am still seeing the same approach in M&A PMI projects, the big mistake being the assumption that technology implementation will address what is primarily a people problem.

People are at the core of these projects, and the impact goes far beyond the office. It affects the community, suppliers, and individuals like the taxi driver. Only by engaging with them will we ensure that they remain motivated during a highly disruptive and turbulent M&A technology project.

It really is that simple – if you want better M&A results, develop better engagement with your technology employees.

What about you?

The secret of your future is hidden in your daily routine.

MIKE MURDOCK

As the core principle of this book is 'people first', the first person you need to look at is yourself. It would be unfair to assess your team and the acquisition without having full awareness of your innate skills.

In the 'Engage' section of this book, I outline how to run conative testing for the team(s) you work with. But before you do this, I highly recommend you run a Kolbe test against yourself. It's even better if you can run another for a colleague you work with frequently, because a mutual understanding can allow you to work with each other's strengths. You can run a Kolbe test either by logging on to www.kolbe.com or by arranging a test through us. My team and I will help you interpret the results. You can find more details from http://info.beyond-ma.com/BeYourself

I appreciate that I am asking you to invest time doing something different, but if you want to break the mould of failed M&A projects and you want a happier, more engaged team, your approach must be different from the outset. Armed with the information from the Kolbe assessment, you can be free to be yourself, which is something worth fighting for.

You can also receive a free 30-page report assessing your post-merger integration risks by going to http://scorecard.beyond-ma.com/

Warning: previous M&A experience may hinder the project

Opportunity is missed by
most people because it is dressed
in overalls and looks like work.

THOMAS A. EDISON

Have you ever been camping in thick snow? I've been lucky enough to do it once, down on the South Island of New Zealand, and it went well. I remember pulling the zip of the tent down and seeing snow-capped mountains in front of me. Awesome.

But there's no guarantee camping in snow will go as well next time. We have no control over nature, snow and mountains, and this is almost certainly the case with PMI projects, too. So it's worth emphasising that having previous M&A/PMI experience does not guarantee repeat success. The multicultural and multi-party nature of these complex projects means that no two PMIs are ever the same.

PMI projects can be stressful, take up a lot of your professional energy, and perhaps many years of your career. Therefore, carefully consider your overall approach to each project so that your skills and experience work to the benefit of both you and your customers.

I suggest four principles to guide your approach:

Design the team, not the technology. Every project I've worked on spends more time in the initial stages focusing on the technology design. This approach will generally confuse everyone and make the work seem insurmountable. Even directing a tiny proportion of the time, money, meetings and workshops towards actually designing the team will give you a much better outcome.

Place people in the right positions based on their innate strengths – which you can rapidly discover, using psychometric assessments.

Keep an open mind. Bringing new businesses together will create or highlight unique issues that you will probably never see again. Aim to identify those issues as soon as possible to determine how you can help your team respond to and manage them.

Promote genuine creativity. Pause the doing and encourage thinking. PMI projects generally throw everyone in at the deep end, meaning a lot of doing happens before anyone's had time to consider what the impacts of the actions are. People need time to think, be themselves, and be genuinely creative. Meetings and workshops are not the spaces where this happens.

By keeping an open mind, you can help your tech team to 'open the tent door' and be in awe of what they see in front of them.

PART ONE

The Challenge Of PMI

Chapter 1

What Is Post-merger Integration?

Some people don't like change,
but you need to embrace change
if the alternative is disaster.

ELON MUSK

Some time ago, I worked in New York City as part of a disaster recovery project. From the day I arrived, I had amazing experiences living and working in Manhattan for an investment bank, making some good mates. But a year after I arrived, my gut was telling me to do something else.

As my project came to an end, I was offered a good position which I turned down. My manager, a larger-than-life New Yorker, was totally alarmed with my choice. When he asked me why I wanted to leave, I had no concrete reason. I loved my NYC experience but I just had the urge to leave. I had an overwhelming need to get back to London, even without a job.

When I returned to London, the adjustment took a couple of months and was difficult. I started to doubt my decision to move back home – maybe my boss had been right.

But nine months later, 9/11 happened. In New York, I had been working in the Twin Towers. Of course, I was very lucky not to be there on that terrible day.

You need to listen to your instincts. And your instincts are likely to be firing almost daily on a PMI project, because so many people are involved and so much change is happening. Sometimes you'll need to be brave to react to your gut, especially if it's telling you to do something unexpected.

Mergers and acquisitions, or M&As, are well-known to many. They are often announced and covered in the national and international press and are studied in-depth at universities and business schools. In my opinion, that's not surprising considering the incredible impact they can have.

Once the M&A deal has been agreed, that's when post-merger integration begins. But what is it, exactly?

> *Post-merger integration, or PMI, is a complex process of combining and rearranging businesses to materialize potential efficiencies and synergies that usually motivate mergers and acquisitions. The PMI is a critical aspect of mergers; it involves combining the original socio-technical systems of the merging organisations into one newly combined system.*
>
> Wikipedia

> *To capitalise on the latent synergies of the newly merged or 'transition' organisation, the transition process itself must reflect the principles of – and provide the conditions for – synergy throughout its design and execution. This means promptly involving people from both organisations in core business tasks (selling, planning, decision-making), as well as on merger project teams. It also means visibly demonstrating a commitment to learning and to creating*

> *together something greater than either party could create on its own.*
>
> Institute for Mergers,
> Acquisitions and Alliances (IMAA)

Both explanations indicate complexity, but for me personally, the phrase 'combining the original socio-technical systems' (of two or more companies) nicely summarises PMI. There is a strong relationship between the technological and the social challenge, and both need to be addressed in order to deliver the integration well. You can have great tech delivery and systems, but if the staff aren't fully on board, there will be issues.

Merger, acquisition or divestment?

The type of PMI project will be related to the type of M&A activity that has taken place. From a technology perspective, the integration activity related to a merger or acquisition will be similar: bring two or more technology environments together. It's worth remembering, though, that a merger is rarely undertaken by two equal parties: there is always one dominant party, and this will have an impact on your delivery.

By contrast, in a divestment project, you need to look at separating business unit and the associated technology services from the main business. This can be challenging when it comes to shared computer systems, data and information, i.e. when a system will be used by both the separated business and the business that remains. Depending on the service involved, your team will need to determine who owns the data and whether all, or any, of the data can be moved into the new company.

A combination of the two situations is also common: where a business unit is carved out from one business and then merged into the purchaser's business.

From a board perspective, the acquisition appears simple: either merge or divide company technology assets and ensure there is a financial benefit. In fact, I often meet people who are involved with the business acquisition, but have no idea of what happens after the deal. But for the senior technology management team, the transformation is a headache involving complex projects and a ton of workarounds that are often enterprise-wide and spread all over the world. It is essential to understand why the M&A activity is occurring and what the benefits, other than growth, are likely to be. Once you understand this, you need to communicate it to the wider team if you are to get the best performance from them.

There is plenty of information available online to help you understand M&A drivers. Here are some of the more common ones.

Mergers and acquisitions are usually financially driven, the most often quoted reason being to create synergy. In other words, they create value by opening up a better and wider customer base, and helping the company to grow through market expansion or entry into new markets. The eBay and PayPal case study is one of the most successful synergies and a great example of a perfect match.

The acquired company may have different but complementary technologies, skills, patents, intellectual property or employees which will benefit the acquiring firm. Or maybe the acquisition is defensive, to offset threat and increase future bargaining power.

The key is to understand the reasoning, the intended benefit, and relay this back to the technology team. Starting off with a transparent purpose and mission will help to create a more inclusive environment and coherent team.

A company divestment, or carve-out, can have a big impact not only on the operations of the subsidiary being divested, but also on the financial stability of the entire business. This can affect how the technology staff view the change to their jobs.

The reason for divestment may be that the company needs to find cash to fund future investment requirements. Or it may be the business unit is no longer part of the core business, or is currently weak in the market. Maybe the divestment is a part of a future strategic partnership, which could represent an opportunity for the technology staff.

Spend time identifying and understanding the reasons for the divestment to ensure employees' reactions and expectations are aligned with the planned changes ahead.

People first

M&A PMI projects are complex. Regardless of whether the acquired or divested business is large or small, there is a need to assess and plan highly complex system changes while 'keeping the lights on'. However, there is a temptation to focus on the project management and technology rather than the most important factor – the one that will decide success or failure. The people.

For example, let's take a simple acquisition of a small business of a few hundred people by a large business, and assess it purely from a technology standpoint. When a business is acquired, the executives at both companies are aware of the deal long before it is announced, yet the upper management of the technology team are often left in the dark until it's made public.

Both the buy-side and target-side businesses will have their own culture and unique methods of delivering technology. As part of the delivery, the two organisations will accept different risks. And both parties will prefer to keep the reality of how they run technology change management to themselves. Everyone takes shortcuts and creates workarounds which are not of any concern until an external party assesses those systems and processes.

We cannot make assumptions about how two cultures will work together. When we assess the above, most of the challenges have very little to do with technology itself. They are more about how people operate their technology and how they will interact with others outside their organisation.

Simply put, an acquisition requires people representing or working under two or more parties to share information that they would prefer to keep within their own organisation. This scenario will be the foundation of every single PMI project you will work on.

Most likely, technology staff from both sides, who will be the ones to make the transformation happen, will find out about the acquisition via a press release.

They may only receive limited information on the M&A deal benefits. An acquisition is intended to further the growth of the buy-side company, but what does that mean in practice?

If you are representing the buy-side company, you will no doubt meet many people potentially impacted by the deal. Beneath the surface-level concerns about the technology challenge, they are likely to be thinking, *Is my job secure? Will I be able to pay my bills? How will this affect my family?*

Both sides will need to reveal how they operate. Generally, the practices they use will have been built upon industry and company culture, organic growth, and what worked at the time. Having to reveal these internal practices can be a challenge, but being open is essential.

For a tech employee, the M&A deal can be a positive or a negative, depending on their role, experience, skillset and willingness to learn. They may embrace the challenge, or be fearful of the upcoming change. Therefore, to deliver the best possible PMI, you as a leader must focus on 'people first'.

Not only are staff likely to be concerned about their job security, they will also often have been signed up to a project timescale that they have had no say in. Sometimes they won't believe it is feasible. Sometimes an aggressive timeline can actually jolt employees into action, and they may even enjoy the experience. It takes time and effort to get genuine buy-in from the tech team as a whole, but this is essential if you want to get the best performance from them.

Lastly, it is likely that technology skills will be assessed throughout the entire delivery in order to see how staff will fit into the new combined business. Therefore, you need to lead the team, engage them, and help them deliver work of the highest level that they can achieve. That way, they may end up with better roles in the new world.

IT due diligence

It is often quoted in the media that 70–90% of M&A deals are unsuccessful – as in, they don't return the assumed or intended benefits. As a rate of failure, that is simply sensational. So why do something with such a high likelihood of failure?

It's human nature to grow and take risks. For example, only 2% of films make money (which in Hollywood can be down to creative company formation and accounting), but there's still plenty of investment in an industry with a 98% failure rate.

When we drill down into the detail, the following reason is often given for the high rate of failure of M&A deals: 'The new company formed (or divested) by the integration project did not provide the intended value to the shareholders' or 'The intended synergy didn't work because the technology failed in some way.' But if we dig just a little deeper, we find the reason is usually a problem with the initial due diligence. Which essentially means things were not as they seemed.

If technology systems and services are not fully discovered and understood, the resulting integration project will be majorly impacted. We can only plan for what we know we need to change. To make matters worse, many technology environments have grown organically over years, yet the PMI team is expected to uncover and understand everything in a matter of weeks.

Why is IT due diligence a key part of M&A PMI programmes? People may be reluctant to disclose key information that will transform your project, and it doesn't help when they are pressed for this key information early in the project before they've had time to trust you, your team or your approach.

In addition to information disclosure, finger-pointing between businesses during the pre-deal, R&D, or implementation stages can further complicate due diligence. Therefore, if rapport can be built between teams from both sides in the early due diligence stages, this will be conducive to a smooth change process and better teamwork when things go wrong. The technology teams must trust each other and feel they can share information.

Later in the book, we will examine why due diligence challenges arise and how they can be resolved.

Leadership

While every change project, IT-related or not, will benefit from a strong leader, a PMI will simply not work without one. By strong, I mean effective, believable, talented, etc. I'm talking about someone who knows what they are doing, and can convince people from different companies and cultures to make a change. While a lot of focus will be on the Chief Information Officer(s) in the deal, who will of course set the boundaries, the real change makers are the programme and project managers.

A strong leader is essential due to the various needs of multiple companies and multiple stakeholders. The leader must have the ability to convince wide-ranging people, from the project board to the engineers, that certain changes and risks are necessary to get the project over the line.

Summary

Due to the complexity of a PMI project, there can be a temptation to make each activity a technically focused endeavour, when in fact most of the challenges are people-related, in particular the changes people are facing and perhaps fearing, both personally and professionally. While human resources may address general career and culture changes for the entire businesses, they are unlikely to be looking at the technology-specific challenges of the affected tech professionals, and how their performance during the transformation will impact on how they are perceived in the new company.

I advise you to analyse and plan an M&A PMI transformation in two phases: one that looks at the people, and the other at the processes that will enable delivery. Part Two of this book will discuss each of these phases, but before we move on, let's understand how M&A PMI projects differ from standard technology projects.

And always, like with my New York experience, regardless of any framework, remember to trust your instincts.

Chapter 2

The Challenge

If you think you can do a thing or think you can't do a thing, you're right.

HENRY FORD

Indonesia. My wife and I were celebrating our honeymoon and we loved the island. There was so much to do and it was the perfect place to focus on ourselves following the build-up to the wedding.

We noticed there was a volcano climb available from a local tour operator, so we booked it and joined the excursion early the next morning. When we turned up, everyone else was wearing the proper gear – tough shoes, hats and walking sticks. We'd turned up in beach clothes and flimsy trainers, and I personally didn't have anything to cover my head from the blistering sun. I'm pretty fit, but once we got going, even just walking without scaling the volcano was exhausting. I wasn't sure we'd last the distance.

Eventually we found our pace and spent four days with our guide, climbing up and down the crater and staying in jungle areas.

The traditional moral of this story for us should be 'be prepared'. But in reality, it was the guide who was prepared – and that's the type of person you are on the lookout for to support your PMI project.

How PMI projects differ from traditional IT projects

In order to plan more effectively and deliver a better project, be well prepared and have the relevant teams on your side. To enable you to do this, it is important to understand the characteristics of PMI projects that make them unique. This chapter will outline the five distinguishing characteristics of a PMI project:

- Multiple parties
- Amplification of traditional project concerns
- Public awareness
- Competing change control mechanisms
- Project slippage

While these components look similar to a normal tech project, the inherent complexities, scope and relationship challenges are amplified in comparison.

We will now discuss these challenges in detail.

Multiple parties

Once a PMI project has commenced, there will be pressure to deliver Day One integration as soon as possible, often on a pre-determined date. Therefore, it is wise to build a PMI plan that incorporates as much team-wide co-operation and acceptance as possible.

In its simplest form, a traditional technology project structure can be operated by two parties – a single department delivering new technology to a single business/unit. The technology department may also be implementing a single technology or service (although most will need some form of integration with other business services). In addition, a traditional transformation is usually performed internally with the output consumed by limited stakeholders. The main benefit is simplification of communication and change.

However, a PMI is a different story. It will involve a minimum of two sets of business stakeholders and two technology teams. Along with those teams come internal processes, environments, cultures, risk appetites and overall approaches to IT service management, all of which will be subject to scrutiny post-deal. Among the different parties there will be high expectations of the others' technical capabilities, but technical capability isn't the benchmark to measure when it comes to delivering technology. It's about how the team operates as a whole.

In a PMI project:

Each party will protect their business technology services first and foremost. However, gridlocks can be reached if neither party is prepared to be flexible to enable progress. To avoid or break a gridlock, you need people on your side who can drive and orchestrate change.

Any new idea or approach must be assessed and approved by two or more teams. As an experienced agent of change, I'm sure you are already aware how difficult that is with just a single technical team.

Additional verification and vetting can slow down every single change. Multi-party assessment and authorisation can erode your chances of meeting projected deadlines.

There can be a lack of accountability. Responsibility to deliver can shift back and forth between different parties. This is especially true if each party is managing or administering different aspects of the same system(s).

Excessive vetting processes can slow momentum, which in turn can take the energy and fun (yes, it's there if you want to find it) out of a potentially exciting and transformational project.

Failure of riskier changes can create a 'told you so' attitude. No one appreciates hindsight driven by negativity, but this type of attitude can become prevalent if your teams are not collaborating well.

Any party can derail a project. Remember that the PMI activity is not part of business as usual (BAU) for either side so will not be a priority for production service management teams.

Ultimately, a PMI is an unfathomably complex socio-technical environment, especially if you are managing a multi-party team where your neck is on the line.

If the impact of multiple parties has not been assessed or catered for early in the PMI project, the following may happen:

NOTES FROM THE FIELD

Them and us

In a recent project I worked on, the buy-side company was acquiring a business of a few hundred employees. There were some early assumptions made in regards to how long the integration would take as the business to be acquired was small and the larger company had the resources to make things happen quickly.

It transpired that the larger corporation had a much higher appetite for risk when it came to technology upgrades than the small company. But there was no formal assessment of either team or how they operated before they started to work together.

To save money and time, the larger business wanted to adapt the acquired systems on to newer technology platforms. They were happy to upgrade operating systems and application platforms and take on the post-implementation support, confident the use of new software would not be a problem. However, the smaller business refused to buy into the proposed change. And they challenged every proposal that was made, knowing full well that they could rely on ITIL service management best practices and vendor documentation to back their argument. The corporation could not argue against best practices so they were forced to absorb and re-introduce older technology that they had previously eradicated from their environment.

To make matters worse, the acquired business reported what they perceived as 'reckless' use of new technology to the joint project board. Creating a fuss at this level rattles many cages and makes thing appear more dramatic than they really are, which leads to further assessments and delays.

The teams continued to work in this fashion throughout the integration, creating a 'them and us' scenario. This made for a hostile working environment which persisted long after the Day One integration.

Even in a two-party project, the differences between the teams created delays and additional costs. How can you address the challenges posed by the multiple parties involved? I recommend you spend a short time assessing the team members and change control processes before starting any project work. Speak to the appropriate people on each side to determine how they deal with change and where they think the potential issues will be in the future.

It is worth sitting on a couple of change acceptance board meetings with each party in order to assess their effectiveness and identify potential delays. Determine the average time it takes each party to make a change, any lead times or risks your teams may need to factor in, and observe proceedings and leadership. You'll get great insight into anything that may impact the work you plan.

Due to the multiple parties involved in a PMI project, be the change big or small, a communications and stakeholder map needs to be developed, taking into consideration each party's preferred communication style. The objective here is to eliminate assumptions. By using this map, you can determine the correct people to consult when you're making a change and thus avoid any unexpected reactions.

Once you have information about the change processes and have developed a stakeholder map, you will have a high-level view of how the project will operate. But due to the nature of a PMI, it is worth going one step deeper and assessing the technology teams. Assess them formally, before any project activity is started, using personality, conative and technology profiling tools.

This assessment is likely to uncover that some of the people in the team(s) you have been asked to manage are in the wrong jobs (e.g. an introvert is in a leadership position or a go getter is working in support). Contrasts between an employee's personality and current job position can deliver poor results. Upfront psychometric profiling can prevent expensive mistakes because it gives you much richer information with which to formulate a plan.

Amplification of traditional project concerns

Regardless of the type of PMI project you are managing, you need to be prepared for every aspect of the project to be amplified. This simply means that the cultural and people challenges are far more complicated than usual, so the element of surprise and the risk and impact of failure are higher. This can lead to increased levels of anxiety and stress across the team as they learn how to deal with a complex environment that is continuously changing. Being prepared for project amplification will enable you to ensure the project doesn't suddenly become unmanageable, as you will already have measures in place to avoid things becoming overwhelming.

On a positive note, once a PMI project has been completed, the sense of achievement is usually high, and there is the opportunity for the team members to gain tangible benefits by learning new technology integration, project and people skills. If the technology integration is successful, it may lead to greater security for the business and staff, and many more acquisitions and integrations. It is important to remember, though, that one successful integration does not automatically mean another will be just as successful.

Public awareness

Another key feature of a PMI project is the public awareness of the Day One deadline. In this context, the 'public' may include:

- The joint project board
- The shareholders

- Numerous technology teams
- Technology suppliers
- The general public, by way of national and international press

The project deadline will be set early and will be determined by factors outside the IT department's control. This date has a major potential benefit in that it can drive the joint project team to work at an extremely fast pace, which in turn creates momentum. However, speed and pressure do not always lead to the best results. They can exhaust your staff and lead to mistakes or failure further down the line. Within the project itself, you need to link the technology project to all of the business changes taking place at the same time, from logistical changes, such as office moves and fit-outs, to more complicated changes in business processes, re-branding and teaching staff how to do things differently once they go live.

If any slippage does occur, it will have a major impact on the entire team and process, so be prepared to face a deep level of scrutiny and excessive meetings to establish why slippages have happened and to convince the board that they won't happen again.

Let's explore some real-life examples where a public deadline has caused issues.

NOTES FROM THE FIELD

Lack of input from the people who matter

In one integration project, the technology project schedule was determined with little to no involvement of the technical teams. The project manager did not build the plan collaboratively. Instead, he built a high-level plan on his own.

When the plan was presented to the steering board, they were happy as it had been shoe-horned to fit a date that was acceptable to them. However, the plan did not take into account the time needed by the technology teams to deliver. This created resentment and significant concern in regard to the overall success and quality of the integration. Notably, this resulted in numerous project deadline extensions, a disgruntled project board, and the eventual dismissal of the project manager.

The project was eventually re-forecast, adding significant time and budgetary requirements. Worse still, the entire team was then working on the back foot, seen by the board as always being behind. But this was actually because the original forecasts were never realistic. This kind of situation creates significant pressure, negativity and stress.

The lesson? Ensure the team helps build the plan, otherwise they are likely to reject it.

While speaking to engineers etc. for estimations and factoring-in their delivery time might seem like basic project management, it is important to highlight how many people will be impacted by a poor approach. It takes time to rebuild bridges and confidence within and between teams.

In this next example, the public deadline caused tangible issues – namely, people became unwell due to the pressure:

NOTES FROM THE FIELD

A grudge

I met with a project team leading a technology divestment. Differences between how the management and the engineers had understood the plan had created a deep divide within the team.

In theory, this project should have been easy to manage as everybody already knew each other. The project manager had developed a detailed plan with thousands of tasks, and the deadline had been made public, but the engineers did not agree with the timescales. More worryingly, people had started to take time off sick due to stress, and because they disagreed with the schedule, some contractors left the business. The technology staff, who were crucial to the integration, were effectively holding a grudge, and without them on board the project was doomed.

There were three views on when this project would complete: the view of the investors, of the project manager, and of the people doing the work. No one could come to an agreement, and a year had already been spent in this deadlock. The managers had been focusing on new technology, not the end goal. As a result, the management team was replaced and the new team given the remit to re-develop the plan from scratch, and re-engage the engineers.

This proved to be an uphill struggle. The new plan had less granular detail than the original in order to make it feel achievable. The team broke the programme into three different phases over twenty individual work streams, each with an accountable owner. As it felt more realistic and manageable, the engineers started to buy into the project and deliver. We were able to re-mobilise the project and make an impact within days, breaking down the barriers to change.

The lesson: involve the team in the planning stage. It's an expensive exercise if you don't.

Unlike a traditional IT project, a PMI often feels like the deadlines are set in stone and that slippage simply cannot occur. However, you need to fight for a realistic timeline in order to get buy-in from technology staff.

Competing change control mechanisms

Change advisory boards (CABs) are necessary within enterprise technology. They aim to bring representatives from both the business and technology teams together to discuss change from their different perspectives, investigating the impact on the business and securing the company from harmful change.

While the Information Technology Infrastructure Library (ITIL) provides industry guidelines on how to run a CAB, each business will operate them to different degrees of detail and success. A CAB is very much aligned to company culture, the industry the company operates in, and its appetite for risk. How the CAB is operated, the time it takes to deliver change, and the overall effectiveness of CAB processes are specific to each business. To some, a CAB is the cause of delays because it's a documentation process rather than a method to implement future proposed changes safely.

Consider a PMI project. Bringing together two or more CABs from different businesses will almost certainly introduce further complication and delays. Sometimes, the CAB processes will clash because (1) one party agrees to a modification that ruffles the feathers of the other CAB, which then refuses to make the modification, or (2) one CAB is slow to respond which impacts the other companies.

Early awareness of CAB processes will help you to deliver a more accurate plan. Learn about the processes from both sides by:

- Documenting what systems are used to record and implement change

- Asking who the key players within IT and the business are
- Asking how long it takes to deliver change
- Discovering the rate of rejection – i.e. how effective is the CAB? Is it a documentation exercise?
- Producing a schedule of change freezes that may impact your project

Awareness of key or seasonal business activity is essential, for example a seasonal sale for a retail business, or a specific sporting event for a betting firm. The PMI team will lose credibility by proposing change during a known period of critical business activity, as it demonstrates lack of knowledge of their business or lack of interest.

Change control during an acquisition. To help identify any future risks that may delay your project, I recommend you assess the CAB processes of each company. Make sure there are mechanisms in place to inform each CAB of upcoming change and make them more aware of your project. Speak with the CAB administrators to identify any problems that may be encountered.

Getting CAB administrators on your side is easier to do before you start making changes. Once you commence with the changes, your team may hit unexpected difficulties (technical or otherwise) which may mean the CAB administrators become more restrictive or allow less free rein. Meet with them and help them understand the importance of your team's work and the impact on each company and employee before making changes.

Sometimes, although it's not deemed industry practice, it is possible to ask for a 'blanket' change which would allow all your changes to be made under a single approved CAB request. Or consider putting many changes forward for assessment and approval in batches. However, often companies dislike blanket changes as there is the risk that undesirable changes could be slipped under the same request. But it may be worth pursuing, depending on your appetite for risk and ability to control the changes.

Change control during a divestment. The complexity of change can vary between projects that come under the umbrella of divestment. A divestment can be as simple as breaking data centre and networking links between two entities, or as complex as setting up a brand new company with its own IT systems which needs to co-exist with the original business until the transition is complete. It is likely that the company or business unit being carved out will not have its own CAB board. Therefore, it is the PMI leader's job, on behalf of the new business, to ensure that the correct level of detail is considered and care taken.

Even if you know the company's CAB processes intimately, it is essential that you assess and record them with PMI requirements in mind.

Project slippage

Compared to traditional technology projects, PMI projects are more complicated. The stakes, the need to succeed, and the reliance on IT to make things happen are higher. These factors create a far more challenging project landscape as the projects generally rely on a high degree of assumption throughout.

Building a project plan to deliver on the Day One integration date is essential to maintain confidence across all parties. Unfortunately, few projects manage to progress according to the original forecast, and increasing the project timeline may turn the endeavour into a negative one as the team feels it is never on track. The situation becomes even more pressurised if the financial implication of project slippage is high.

There are five main causes of project slippage:

- A Day One date being established without consulting the team
- Assumptions about technology and the transformation timelines during the project planning stages
- Late identification of a service or system
- Lack of detailed information
- Lack of technology team buy-in/engagement

If a project extension is inevitable, identifying this need and dealing with it early may help give the team some space to breathe. Many PMI projects can overwork employees, which leads to unexpected absence and sickness. Providing some slack helps to avoid these problems and facilitate a smoother project delivery.

Summary

In Chapter 2, we covered how a PMI project differs from a standard information technology project. The following list highlights some key considerations for your next PMI project:

- The duration will vary, with some PMI programmes taking many years
- Depending on the overall M&A deal, some of the IT staff may feel disoriented, uncertain, or reluctant to work
- The risk and rate of project failure is high, so be sure to discuss this at project initiation
- The impact of IT on the overall M&A deal, both financially and in regard to achieving a better service for customers, is high
- Make the staff aware of the why, i.e. the benefit to the company and the potential benefit to them personally
- At least two IT departments and two businesses will need to work together
- The projects are generally complex because the team is transforming a full IT environment
- Once the Day One deadline is made public, it is scrutinised by many parties and difficult to change
- Problems cannot be hidden by the IT department, as another IT department will be assessing its activities and change
- At least two CABs will be operating, each with its own processes and inherent delays

Overall, due to the number of parties involved, everything in a PMI project is amplified, which returns to my fundamental guiding principle: PMI projects are people challenges, not technical ones.

Chapter 3

Potential Risks

Fate loves the fearless.

JAMES RUSSELL LOWELL

Mid-career, I decided to go to university and managed to get a BA in film and an MA in screenwriting – a bit of a detour from technology, but I felt I needed to widen my horizons. When I finished my studies, I decided to write a film about a road-trip across India, and thought it'd be more genuine if I could actually do the motorbike ride myself. The only problem was, I'd never ridden a motorbike.

So I took a one-day 'CBT' course in London to learn how to ride a scooter and managed to persuade the Indian embassy that this was sufficient to ride a full-sized Enfield in Chennai. I remember that feeling of pure fear when I was 'handed the reigns' by the mechanic. It was so damn... *heavy*. While I am not one to map every move in detail, I was aware of the risks – India is the most dangerous place in the world to drive, and I didn't speak the language. So I took my time initially, getting practice riding the bike slowly around paddy fields.

But even with known risks, new things will crop up, and India's worst monsoon for thirty years hit the region: the water was up to my waist. I clearly remember pushing the bike through an expanse of water known to be diseased, wondering why I'd chosen this endeavour.

But what struck me was how others took this in their stride: this was a major event for me personally, but life carried on for the local people. So, what may appear as a risk to you may not be a concern for others, so while it's important to identify potential risks, it's also essential to reality-check them with others and find people in your PMI team who can help push past them and keep the momentum going.

Workarounds

You may have come across unplanned systems that end up growing into an enterprise-wide necessity. These in-house systems often start off as small fixes, or workarounds.

Handled incorrectly, workarounds create technical debt: the implied cost of rework caused by choosing an easy solution instead of a better approach that would take longer to implement, often with no plan to remove the workaround in the future. For example, a workaround may create an increasing requirement for computer resources that is not obvious to management until the costs are seen, long after the PMI project is over.

In a PMI project, workarounds may be put in place to avoid a public project extension and maintain the board's confidence. Workarounds are often informal and lack documentation. I recommend that you record any workaround you use and develop a remediation plan, with agreed actions and budget, to reverse it. Include resourcing and any associated costs, such as licensing, in the plan. Not only does this approach mean a record is kept of the Day Two actions, it also ensures any associated costs do not come as a surprise once the company has been fully integrated.

NOTES FROM THE FIELD

Record the workarounds

A client was moving a corporate IT infrastructure from one managed service provider to another as part of an M&A deal. During the project, it became apparent that the underlying technology - which was a major reason why the new supplier had won the contract - did not work. Even though the technology was a well-known solution that had been commercially available for over a year, the teams had little real-world experience of the software, and it did not meet the business needs as promised.

To meet Day One requirements, the computer systems were 'lifted and shifted' across with none of the promised new functionality, security protection or insights. The problem here was that the entire solution was a workaround. Making changes in the future to bring the system up to the intended design would be difficult because of the way in which the business operated. The stakeholders would accept disruption during the initial transition, but not beyond.

While this is also an example of due diligence failure, the main problem was the failure to test the proposed system. This is a common occurrence in PMI projects. In this extreme case, where the entire solution was a workaround, the company needed to rip it all out and start again. This is a poor foundation to build upon.

The lesson: keep your eye out for tech workarounds and make sure they are recorded.

Keeping track of workarounds is essential as they need to be reported and maintained long after the integration has completed. This is especially important as the PMI team generally disbands soon after the Day One integration. If no one is aware of a workaround and why it has been put in place, they may be nervous about changing or removing it. At the very least, workarounds need to be listed in the post-implementation service review and assessed for cost and effort to remove them.

Security

Be careful that your project does not create a security concern. It is estimated a whopping 40% of PMI projects introduce a security issue, which is another sensational statistic. It is important to think about the legacy that you will leave behind, even though this can be difficult to do when you're under pressure to deliver.

A recent report by West Monroe Partners[1] found that:

- The cybersecurity issues of 80% of respondents had become highly important in the M&A due diligence process
- 40% of acquirers had discovered a cybersecurity problem at an acquisition after the deal had gone through, indicating that due diligence standards were low
- Compliance problems were one of the most common types of cybersecurity issues uncovered during the due diligence phase of 70% of respondents

---

1 www.westmonroepartners.com/Insights/White-Papers/security-survey

- A lack of comprehensive security architecture was a problem for 40% of respondents

You can imagine the difficulty of working with two or more chief information security officers, each putting forward their own standards and security agendas. Security is a sensitive area right from the initial assessment of an acquisition, during any post-merger planning, and on Day One of the transition, but the main risk comes from security workarounds that were put in place temporarily and were not removed after Day One.

Transitional service agreements (TSAs)

A TSA is often used during a divestment, or by a less sophisticated buyer where senior management is in place, but the back-office infrastructure has not yet been assembled. If a buyer does not have the management, people, or systems in place to absorb the acquisition, the seller can offer them for a fee, but generally make no profit.

TSAs clearly define the following key points:

- The fee for the agreed scope of services, and additional charges for services beyond the scope
- The terms of the agreement and any available renewal terms
- The dispute resolution mechanism to deal with discrepancies between the buyer and seller
- How the TSA will be transitioned once the buyer is able to assume the administrative responsibilities

TSAs can be extremely difficult to manage if they are not properly defined. Poorly drafted TSAs usually result in disputes between the buyer and the seller concerning the scope of the services to be provided.

Use of business-as-usual staff for PMI project delivery

This is an extremely common scenario. Business-as-usual (BAU) or service support staff are usually maxed-out supporting the current business needs, but they are often utilised to help build and deliver the integration project, too. The main benefit of this is that they know the people, systems and IT environment intimately, and understand the impact of change. However, if BAU staff are expected to deliver project work in addition to their day-to-day work, it can cause major delays due to one or more of the following reasons:

- They may be de-motivated
- They often cannot dedicate time to delivering the project
- Generally, they are not used to complex company-wide IT integration projects
- They are not seasoned project consultants
- They may not have the drive to push the project delivery at the level required
- They may not work the extra hours required
- They may not have an impartial view of the systems
- They are unlikely to be ambassadors to sell the change

It is likely that budget constraints will mean that the BAU team cannot be backfilled. Therefore, project managers and technical staff from previous projects help augment, guide, and get the best performance from the BAU staff.

The decision of whether to use BAU staff for a project will inevitably come down to cost and company culture, but you will need to highlight the risks of utilising them early on. Often, using BAU staff for these projects cannot be avoided, so it is worth considering them in your plan and adding some contingency. If your team runs conative profiles on the BAU staff, as I highly recommend, you will gain a quick insight into whom to place into which positions. We will cover conative profiles in Chapter 6.

Technical navel gazing

Can you recall a situation where technical staff have declared that something is not possible, only to implement it the next day? In my eye, that's technical navel gazing. This scenario is a win-win for the technical staff. They attract attention during the meeting where they paint a picture of the impossible, then glory when they seem to have delivered the impossible.

In a PMI, technical navel gazing can be obstructive.

NOTES FROM THE FIELD

A case of navel gazing

A client acquired a medium-sized software business with over twenty years' worth of legacy infrastructure and code that needed to be migrated. The technical staff from the acquisition company were the experts on the bespoke software stack and provided significant input in regard to how the platform worked and would need to be integrated into the new business. Often, they made the platform appear difficult and complicated to migrate, highlighting all the reasons why any project we would undertake would fail. Our team noticed that everything seemed to be 'impossible', but once overtime was allowed these insurmountable problems often disappeared or were fixed within hours. Now we had a real-life example to negotiate with when the team began to 'navel gaze': if they suggested something was impossible, we could show them how they had managed an impossible situation previously.

When you are not part of a tech team's company, to an extent you are in their hands and have to trust their judgement. There are a couple of ways to address this. One is to involve other staff, face-to-face if possible, so that you have more than one opinion. The other is to use someone from your company who has both the technical and communication skills to cut through the crap.

It is essential to nip this behaviour in the bud. For the project to succeed, the technical staff need to focus on the delivery, not their own prowess.

Summary

In Chapter 3, we have covered some risks that are prevalent in PMI projects. I deliberately did not place the 'people engagement' risk in this list as I hope that the message is clear: if the tech employees are not engaged the whole exercise is fraught with risk.

- Technology workarounds made to complete the PMI project could cost the company dearly in the long run
- Security can be weakened during a PMI project
- Transitional service agreements can create long-term problems if not setup-correctly
- Try not to use BAU staff for the PMI project without adequate support and resource
- Look out for tech employees who demonstrate 'technical naval gazing'; this behaviour is obstructive

(Back to my motorbike adventure: I managed to ride hundreds of miles down the south-east coast. The resulting screenplay wasn't that good but I was happy to have witnessed how people can take such difficult situations in their stride. That really put things into perspective.)

Chapter 4

What Tech Staff Are Facing Today

I suppose leadership at one time meant muscles; but today it means getting along with people.

MAHATMA GANDHI

I was taking a connecting flight from Hong Kong to New Zealand. In the queue was a man with bright white hair, and he was certainly a little offbeat. My instincts told me he'd be sitting right next to me.

My instincts were right – he was. It turned out that he was an architect from Paris and work had dried up for him, so he had decided to turn up in Auckland to see if the people there needed any architects. As someone who likes to be spontaneous, I was pretty impressed.

It was a twelve-hour flight, so chit-chat continued, moving to his portfolio of work – interesting – then pictures of his wife naked. Hmm, not so interesting. I quickly diverted to a picture of my white-haired friend and another chap with Bill Clinton. He seemed to be pretty chuffed about this occasion, so I continued to make small talk and asked, 'Where does your friend live?'

Suddenly my new friend became my enemy and pulled a knife on me. As I was sitting next to the window, I had little choice but to stay quiet while he explained his friend was the reincarnation of Christ and that he had to protect him at any cost.

This bizarre occasion always reminds me of what you might uncover when you dig deeper with people. In this chapter, I'll be highlighting some of the factors that tech staff are facing today. Although these factors will not make them carry knives on planes, they will cause some form of pressure for them.

The world has changed

Modern corporations are running multi-generational technology workforces. As technology has become commonplace, so has the technology department, to the point that it has become the subject of a number of hit television shows. Interestingly, most of these shows focus on the dynamics and relationships between the IT team and the wider businesses, and tend to portray awkwardness between people who think and operate differently.

A corporate technology team will comprise a range of different people, from school-leaver tech apprentices to members of Generation X in their forties and fifties, and older Baby Boomers. For the first time, a large proportion of older people have been able to remain at work, using their technical skills to earn a crust. IT staff are enjoying longer technical, hands-on careers and operating large revenue-generating technology systems well into their later years.

Multi-generational tech teams are beneficial for businesses, allowing both modern and legacy ideas, approaches and services to be operated simultaneously. If they're willing, the young can learn from the old, and vice versa.

When a PMI project commences, certain generation-related differences in the team will be revealed, and it is essential for you to understand the people that make up the business. Older technical staff, particularly Generation X employees, are usually capable people who have delivered significant projects to help the business grow. In fact, I would say they've been critical to business growth. Yet I have noticed some seriously worrying trends:

- These staff have been on the IT 'production line' for years and have witnessed and been part of major organisational changes, but little has changed for them personally
- Many have industry qualifications that they gained at the start of their career, but having not changed jobs frequently, they have not kept up to date with certifications
- While many are capable, they have often not been exposed to new environments, soft skills, or technology that would ensure personal career security
- They have far more personal responsibilities than they did when they first started in IT and believe that they no longer have the time to learn
- From their conative profiles, it's clear some have been doing the wrong job for years

Unless they do something proactive, older technical staff are at risk of redundancy post-acquisition. They have a lot of potential to grow and ensure that they are more secure in the IT job market, regardless of whether or not they are kept on. Either way, they will need to learn new skills.

While you cannot personally secure jobs for older employees, you can provide them with evidence that they need to adapt and opportunities within the project to demonstrate that they can and have adapted. In order to engage older staff, you will need to discover their current skills and determine the best course of action for the future. Can you help them focus on work best suited to them so that they can achieve a better position in the new organisation?

Bi-modal IT teams and digital acquisitions

The challenges of multi-generational IT workforces are further complicated by the emergence of bi-modal IT. 'Bi-modal' is a popular buzzword that is often quoted in the industry media, and generally refers to the division between legacy and new technology teams, systems, and services, and, crucially, how they operate differently under the same roof.

There is also a trend for bigger companies to acquire smaller businesses that specialise in online digital offerings. When this occurs, there are often very obvious differences between the parties.

The diagram demonstrates some of the differences between the legacy corporate buyer and a digital acquisition.

DIGITAL	• Individual • Unregulated	• Rapid by Nature	• Per Business Unit • Individual • Self Serve	• Collaborating • ‘Pivoters’ • Less Political • One Person - Many Hats	• Automated /Devops • Innovative	• Own Builds • Public Cloud • Digital
LEGACY	• Controlled • Centralised • Efficient • Volume Based	• Regulated • ‘Slower’	• Tier-1 Banking • Regulated • Risk and Known Liability	• Bigger Teams • Slower • Political	• Appear convoluted • Regulation • Many More	• Legacy Standardised

Underpinning these differences is the time needed to make effective technology changes, which will depend on a company's culture, industry, regulation, and governance. In order to service and maintain both the new and legacy business technology environments, companies may adopt more than one change-making strategy. For example, many companies now find themselves operating a legacy IT services estate running traditional applications and services on virtualised servers, while building their own agile-focused applications and services in private and public cloud environments to meet business needs. These companies often find themselves with more than one approach when it comes to change: a legacy CAB approach and a zippy, agile approach.

Hence, people working in the newer 'digital' environments strive to be more agile. This means actually thinking differently and using a different approach, not just talking about it. The new environment can be changed regularly and at any time, without scrutiny or delay. By contrast, the legacy systems are more closely monitored and are changed far less often.

So what does this mean for the individual when they transition to the new company? It depends on which generation they come from, the type of technology (legacy mainframe all the way up to public cloud and modern micro-services), and which company they represent. For example, if they are in a legacy technology team where it takes some time to suggest and implement a change, they will be used to a certain cadence and (in)efficiency where delays are beneficial because the software solutions don't always work first time. These people need to develop an open mind and look at how they might orchestrate change differently. It is paramount that they speed up and change their approach, but they will need help to make this transition.

In contrast, if staff are part of an agile/modern environment, they are used to rapid change and the perceived business benefit it brings. But if they transition into an environment that mandates a slower change process, this may derail their change processes and the resulting technology solutions. If they become frustrated with the new ways of working, the business risks losing vital, knowledgeable staff.

In summary, the legacy staff need to learn new approaches or technology to avoid becoming victims of transition. And you need to ensure that the new environment does not suffer attrition of the best talent from the acquired business.

One word of warning: you may be surprised when working with digital acquisitions that they are either not as agile as they claim to be, or if they are agile, it comes at a price – namely, a lack of standardisation or controls across their tech services. More information can be found in our whitepaper here: http://info.beyond-ma.com/digital

Orchestrating change

Your PMI transformation team will need to orchestrate change differently depending on how each division or team they interact with operates. For example, you will not be able to request agile-style changes from a legacy team, and if you attempt to, they may denounce your approach as recklessness. This alone is a convincing reason to sit on a few change advisory board meetings, to learn how they operate.

Determine who can champion changes for you and your PMI team from both a legacy and modern perspective, so this is likely to be two different people. Once you have your change champions on your side, you will then need reliable technical subject matter experts (SMEs) to do the work needed to transform and update the systems before being migrated to or adapted for the new organisation.

From the agile team's perspective, will they still be agile in the new environment? Will they experience frustration if they are no longer able to do the job they're used to under new corporate polices? Will your company experience attrition due to staff frustration? Even within a single team, multi-generational and bi-modal differences mean that you have to have multiple approaches. But, to get the full picture, you need to dig deep. Consider how the technology team thinks and how you can help them improve their skills.

Cynical analysis is no longer needed

A major shift in the technology industry is that IT staff now need to think less analytically and more collaboratively. Let's take the system administrator (sysadmin) role. Since the late 1980s, sysadmins have been involved in every IT decision and often had a major controlling stake in change. This created a vast company-wide responsibility, mostly because the software they worked with was often shipped by the vendors with faults. They needed to look cynically at every change and say no before being persuaded to say yes to any updates. As the technology industry grew, these critical, cynical thought processes were praised and valued. And for good reason – they usually protected the business.

In the past, a sysadmin may have helped install a service once and then spent the next three to five years maintaining and upgrading it manually until the next major system upgrade. Add in multiple systems and 24/7 cover, and you've got enough work for an entire team.

In the present day, if a company operates modern systems and services, it is likely to use continuous improvement, or CI. This negates the need to install, upgrade or maintain systems as often as was previously needed. Once implemented, these systems update constantly, making incremental changes on a daily, weekly, or even hourly basis. If a single update does not work, it will be regressed, and no one will be any the wiser. The end result is technology that just works.

With computer systems that are more reliable and updated more frequently, the need for sysadmin is diminishing. Cynical analysis, while still an important requirement to protect businesses, is no longer the only way a technology professional can assess change. Instead, they can work with others and be more social about it.

In this newer way of making change, it's not analysis of the software and if it works we need, but analysis of whether the software works in line with what the business needs. It is the difference between someone who likes to tinker with computers and press buttons and someone who wants to work with people.

With the advent of CI and cloud technology, staff must see new technology and changes positively. To reach their full potential, they must be collaborative and have the ability to become interpreters between services and the business. Those who cannot see themselves working with the business directly will need to learn new cloud technologies, and must consider the likely risk of competing with a younger, cheaper workforce.

Technology is removing jobs, not creating them

Technology has created more jobs than it has destroyed, says 140 years of data.

THE GUARDIAN, 2015

Over the past 100 years, machines and automation have replaced factory workers in mundane and repetitive jobs. This progression was logical, and was driven by a consumer and economic need to reduce costs and increase efficiencies.

In the press, there is plenty of talk about how artificial intelligence will replace humans in white collar office jobs that can be commoditised, such as making decisions on insurance or mortgage applications. Many news reports portray a gloomy future. Yet services like DoNotPay, created by Stanford student Joshua Browder and stated as being 'the world's first robot lawyer', is already going beyond making simple decisions. Through its Facebook messenger app, it has helped 160,000 people overturn parking fines, find emergency housing, or seek and apply for asylum.

Daniel Priestley, the author of four books on entrepreneurship and the cofounder of the Key Person of Influence accelerator, has a perfect slide that highlights the extent to which technology has impacted jobs:

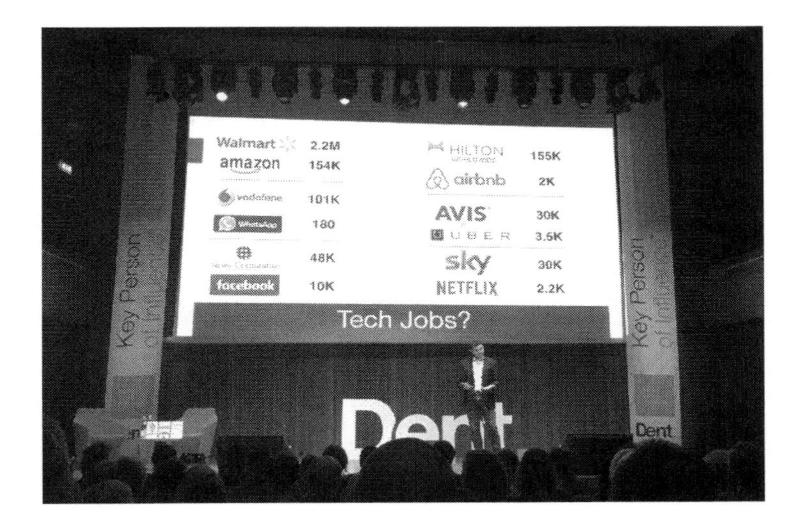

As Daniel reports:

- Airbnb has more rooms available per night than the largest hotel operator worldwide, but it employs just 2,000 people, versus Hilton's 150,000 employees
- Walmart, the largest traditional supermarket, employs 2.2m people, while Amazon employs 154,000 staff
- Vodafone, the telephone operator, employs 101,000 people, whereas WhatsApp, which has a much higher reach worldwide, only employs 180 staff

At the point of a merger, the staff may already feel vulnerable about the security of their current job. But as demonstrated above, the technology landscape is rapidly changing, and they may be vulnerable to losing it for other reasons. Simply put, as computer technology changes, so too does the need for human resource to operate that technology. Hence, there is pressure on employees to change, disrupting their current routines and knowledge base. You need to discover whether tech staff in the target firm have been learning and adapting their knowledge so that they can gain in, rather than become victims of, the new world.

The impact of cloud technology

Cloud technology is an obvious disrupter, and so is referenced frequently in the general and industry media. By running technology centrally, companies are making significant efficiencies in regard to hardware and people requirements.

A great example is email. For over twenty years, all companies employed a resident email services expert, who was paid well to design and deliver a single service. Depending on the size of the company, it might even have operated a global email messaging team. Nowadays, with services like Microsoft's Office 365 and Google's G-Suite for business, email is run, maintained and updated centrally by the software vendors themselves. Businesses simply move all old emails to the chosen vendor and use their public cloud service. There are over 100 million users of Office 365 today, yet Microsoft has a relatively small team of a few hundred people operating the service for the entire world. The team run over 100,000 servers with over 50,000 issues a month; 99% are addressed by artificial intelligence (AI).

Think about that – a few hundred people running a service that in the past would have been operated by tens of thousands across different companies. This means that in most cases, the highly paid email team is no longer required. The people who had these jobs will have moved on to different technologies or roles within the firm, but over time, highly skilled and highly paid jobs have been lost, potentially meaning a loss to the next generation. AI isn't threatening to take jobs in the future; it is happening now. Technology staff need to ensure they have all-round skills and other options in place to prevent any loss of job security.

The good news is there is still a need for people to represent their company as business analysts, infrastructure security and application specialists working with cloud providers, using their knowledge to the advantage of both the company and themselves. There is plenty of opportunity to adapt.

NOTES FROM THE FIELD

The robots are coming

In the 1990s, I'd been working for a global car manufacturer for three years, picking up software development skills. As my career progressed, I was given more responsibility, which meant having to work night shifts. The people I worked with were a bunch of larger-than-life middle-aged characters who drove sports cars and described exotic diving holidays in Thailand – a lifestyle that I could only aspire to, even though they were mostly geezers from Essex, just like me. They maintained their extravagant lifestyles because they were being paid way above the average wage. What's more, they managed to double their wages by ensuring there were opportunities to work overtime after each shift. It was a great deal – who would turn it down?

But the management had other ideas. They gave three of us – the youngest staff – the opportunity to learn, and to change things via automation. Rather than sleep during the less busy periods on our shifts, we learnt to code. We were so motivated because we loved the creativity. The freedom was intoxicating.

Using the coding skills we had gained, we began to automate repetitive tasks. When a problem occurred at night, a computer would address it without the need for human intervention. In other departments, robots took over some of the manual tasks.

We were writing code to replace humans.

Our colleagues soon cottoned on to this and nick-named me the 'Trojan Horse'. I completely understood their concerns, and it was difficult to watch the inevitable carnage. Their lifestyles were suddenly threatened as three teams of fifteen employees were reduced down to five, and then to three, reducing the total headcount from forty-five staff to nine. I can still remember the union meetings, the worry in my colleagues' faces.

The lesson: the AI revolution was happening way before it became a mainstream concern.

How does this story relate to a modern day M&A? While technology has changed, human behaviour and responses have remained the same. My colleagues knew that changes were happening around them as we didn't hide what we were doing, but most of them did nothing to avoid becoming victims of that change. They were genuinely shocked when they were moved into less lucrative positions and lost control of the direction of their careers. The few who did rise to the challenge of adapting were looked down upon by their peers, as the union was trying to maintain the status quo.

How many staff in enterprise technology teams have been sitting in their positions for years with little personal development? Why are self-development books on how to improve technical or IT management skills lying around gathering dust? In this high-tech world, where information is abundant and easily accessible, why do people not pick up new skills to ensure a more secure life for themselves and their family? It could be laziness, but the reason usually has something else underneath it: fear. Fear of change; of having to learn again; of not knowing where to begin.

In the past couple of years, I have interviewed many Generation X technology candidates. The tendency is for older technology employees to gain certain tech skills and then stop learning, becoming a part of the technology production line. The danger is that if a team is on the production line, their jobs are at risk.

When an M&A PMI occurs, there will be an unusual period as the team and their positions are assessed. Are they vital to the business, or are they on the production line? Nobody can hide. The overall objective of the merger, acquisition, or divestment is company growth, so there is certainly a risk for those who are not contributing to that growth. Therefore, it's important for the staff to know whether what they are doing today is vital to company operations or growth.

Possessing or acquiring current technology skills and passion is not age-dependent. I have hired school leavers and Baby Boomers, including one who was old enough to retire, all of whom were up to date with the latest cloud technologies, or willing to learn. A vital characteristic of staff in the new organisation is a thirst for knowledge. Underpinning that, you need the right environmental and organisational factors and support to enable people to learn and develop.

Summary

While I totally believe in a 'people first' approach, I wanted to highlight some of the challenges technology staff may face. Often they are not aware of these challenges until the security of their career is threatened by the sudden announcement of the acquisition of their firm. But the changes are already happening within today's technology industry. You need to discover why it is difficult for people to adapt and change, and how this may impact your project.

If the target technology team is comprised of a multi-generational workforce, there may be distinct divides between legacy and new IT. Older staff may not have needed to learn many new skills in the last twenty years. However, all corporate technology employees are facing these threats today:

- Technology is eroding jobs
- The job market is changing
- Technology is evolving exponentially

The moment the acquisition is announced is usually the exact moment that these threats become a reality for the acquired technology team. Technical people need to be concerned if they are on the production line of the industry. In the long term, it may mean they are at risk of redundancy. And now, it's your job to manage and get the best delivery from them while they feel under threat.

PART TWO

The Managed M&A Method

Chapter 5

Introducing The Managed M&A

A man always has two reasons for doing anything: a good reason and the real reason.

J. P. MORGAN

I spent some time in Goa, India over twenty years ago. My girlfriend at the time and I had travelled with a small group to Palolem beach, which was an untouched cove filled with trees back then. We had fun with the local people and drank feni: a strong spirit made from cashew nuts. There were only two small beach huts that we could stay in. It was idyllic.

As the night progressed, the group became boisterous and dared each other to skinny dip in the sea. My girlfriend wanted nothing to do with it and tried to persuade me not to go, but I wondered when I would I next have the chance and jumped in with the others.

The water was warm and the clear starlit sky looked amazing, and as I moved my hands, a shimmer came to the surface of the sea. This was phosphorescence – an almost psychedelic bright blue light emitted from the water. I've never experienced it again. My girlfriend was disappointed she didn't get to enjoy this experience and I was disappointed for her.

When you get an opportunity, you need to go for it.

Getting teams onside

As we have learnt, people are fearful of losing their jobs for a variety of reasons. It is helpful to alleviate their fears and get them on your side and engaged with the project so that you can make progress together. To engage the staff from the target company team, you will need to help them see the M&A as an opportunity rather than something to fear. In reality, this means ensuring that they recognise the learning opportunities in the situation so that they are able to develop and shine regardless of whether they stay in the firm or not. This approach will help them personally from both a short- and long-term perspective, and it will help your project. If you do this right, everyone can be a winner.

In order for you to help the team help themselves, they will need to trust you and your team. It is essential that you arrive prepared with an overall approach designed specifically for an M&A PMI, otherwise you risk being seen as just another technology team, and in this case one that comes bearing bad news. By utilising an M&A-specific methodology, you will demonstrate both forethought and consideration and kick off with positive momentum.

I have witnessed a few M&A PMI projects where the project managers have not prepared themselves, or the management material, any differently than they would in any other transformation project – and boy, did it show! They started with some project initiation and discovery work, but hit problems within the first weeks because (a) no one believed in them, and (b) they become overwhelmed by the sheer complexity of the task. Without engineer/developer buy-in, they were on their own, and they quickly lost credibility – a critical error.

Therefore, I recommend that you treat the job as an external consulting exercise. One way to prepare is to follow the Managed M&A method, which will help you to pitch, prepare and deliver your project successfully.

Questions you need to ask

Imagine you are the founder/CEO of the target company. It's likely that you will feel the need to safeguard both the company and your employees, as there will be a lot at stake for everyone. You will want to ensure that the acquisition does not proceed until the tech integration approach had been explained, understood, and accepted by your tech staff.

Below are some key questions you need to consider and prepare for. The subtext of these questions is to assure the target technology team management that you can handle such a complex delivery. Have you done this before? Ideally, prove it – demonstrate that you're the right fit. Show that you're a people person, a coach, and leader.

In addition, the target technology team needs to know that:

- You can be trusted
- You and your team are experienced enough to handle this project
- Your technical team are capable
- You have a proven approach that delivers success and won't impact uptime

Question 1: how will you ensure the acquired company won't lose key people? As soon as an M&A deal is announced, people go into survival mode and work out what it means for them. It is human nature. There is a risk that tech staff with key experience and knowledge will decide to leave the company. Seeing colleagues leave can encourage other people to leave, too.

In addition, when the practical work commences, different teams will begin to work together for the first time, and the following will occur:

- The overall working dynamic will change as the team increases in size
- The joint team will span the entire IT hierarchy of at least two companies
- If both buy-side and target-side use consultancies, there will be at least four companies involved
- The technology teams and any consultancies involved will have different agendas
- The combined workforce is likely to be multi-generational and multi-skilled
- The differences in how each team operates and how they deliver technology will not be apparent until they start working together, usually when something goes awry

It will take time to gain the trust of the target company's IT team. While that trust is being built, due diligence will commence, which leads us to the next question.

Question 2: how will you get to know the acquired company's technology environment? IT due diligence is cited as one of the main reasons for M&A deals going sour in terms of return on investment, and brand and reputation damage. A lack of key information can seriously hurt the project. Due diligence can fail because a complex or unsupported service is discovered late, creating delays, adding major complexity, or, worse still, preventing some of the cost-cutting synergy from being implemented.

In some cases, during the assessment you will find out that some of the key intellectual property (IP) that your company purchased does not in fact belong to the acquired business, or it relies on some middle-ware that cannot be transferred to you. This results in risky workarounds and cost savings not being realised as per the plan. Instead of shrinking the team – the intended outcome – you will need to maintain all of the supporting resources and people on each side. These major discoveries often come late and are a great surprise.

Hence, IT due diligence must gather and record as much information as possible within the permitted time. You will need to gather information about software, hardware, finance, support and maintenance, and ensure its readiness for the new world. In addition, early identification and communication of additional risk or R&D work is essential to ensure the available resources and budget are aligned. This doesn't only have a financial impact; finding the right people with the appropriate approach and experience to do the work will take time.

Question 3: how will you create a credible, effective plan?
The project plan and associated technical designs are the basis for the transformation. They need to be accurate, but they must also have as much buy-in as possible.

Early in the process, the stakes are high in regard to the success of the project and the impact on your personal CV. The technology teams may not have been working together for long and there will still be many unknowns, which makes putting a precise plan together impossible. Yet the pressure for you to deliver a plan with associated costs to the board will be high.

The main risk you will face is an incomplete plan or technical design due to lack of co-operation or disclosure from one or more of the technical teams involved. The target company is giving away their crown jewels as part of this project: access to their revenue-generating systems and services. Having to share information that was previously confidential can be a difficult thing for staff to overcome, even if they are now working for the new company. A structured approach will help to convince them to work alongside you, mostly because you will be able to demonstrate in the crucial early stages that you have considered the team and the project as a whole. If you treat the transformation as a traditional IT project, you are likely to struggle as you will need to learn far too much on the job. It may take up every minute of the staff's spare time just to keep up with demand from an unwieldly strategy.

If the plan does not feel right, the joint project team will keep asking for refinement and may lose respect for you as the project/programme manager. A few months running at this haphazard pace will lead to burn-out, and it's not uncommon to witness project managers take a leave of absence due to the pressure of a PMI.

Question 4: can you orchestrate monumental Day One transformational change without impacting the business? All parties need you to achieve the impossible and have everything working on Day One. Therefore, your team will need to demonstrate how they will integrate services in a way that ensures they do not impact the business, nor any other projects and initiatives. It is essential that you answer and document these key questions in the early stages of your project:

- What will the new world look like after the cutover?
- What is the customer goal – how will things change for them?
- What is the technical goal?
- How will the technology team operate in the new world?
- How will the transformation be phased?
- How will the joint delivery team work together?

To answer the above questions convincingly, you need to demonstrate clarity in both the technical design and the project plan. I would recommend documenting the above in a narrative or brochure that will help you articulate and sell the solution to all parties.

Whatever method you use to integrate applications and/or systems within a new company, the impact on the business must be minimal. At the same time, you need to make monumental changes. This is one of the key tensions within a PMI project.

Question 5: how will you maintain and enhance the systems in the future? The customer will want to hear your ideas about how you'll improve the service. It is almost certain that these improvements will have been part of the original reason for the acquisition or divestment. Do your current plans match what the company wants? What are your plans for ensuring a good workable service, and what are your Day Two and Day 100 plans? While the team may not be working with you once the IT systems have been integrated, and they may be long gone before Day Two actions commence, they need to know that you care enough to ensure the current plans have Day Two and Day 100 in mind.

The plan is not just a series of tactical steps to get the project over the line. Ultimately, it is an essential means of demonstrating your experience and wisdom by discussing the real impact of moving and integrating systems, and highlighting the need for additional support post-implementation.

The Managed M&A

The Managed M&A can help you to answer all of the aforementioned questions by providing a standardised solution to a complex puzzle. By using the Managed M&A, you will be able to:

- Get technology staff on your side
- Ensure services and systems are properly discovered
- Create lucid and compelling technical designs and project plans

A PMI project will always be busy and full of curveballs and side tracks. The Managed M&A can help ensure you stay on track and cover all of the important detail while managing uncertainty.

It involves five phases:

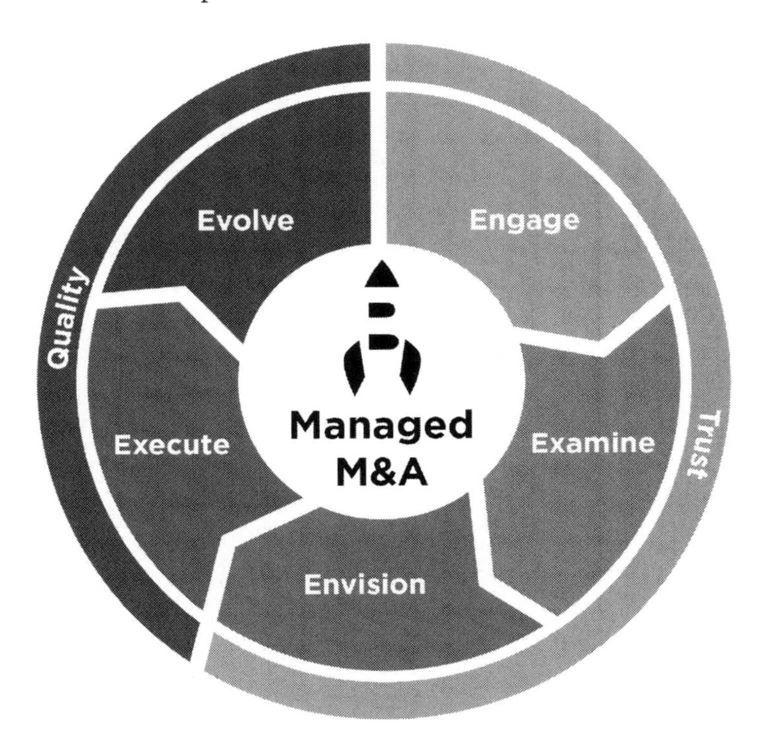

We are going to cover three phases of the Managed M&A process: Engage, Examine and Envision.

Engage. Engaging the technology staff is the most critical requirement of a PMI project. If either company loses staff unintentionally, it will impact the effectiveness of the team and the resulting technology services, which in turn will impact customers.

This section of the book provides a structured way of assessing the acquired organisation, the tech team members, and then the team as a whole. Once you've completed these steps, you'll be able to make the team more effective.

The rush caused by the M&A deal often means that the PMI team gets on with the project immediately, but this results in insufficient on-boarding of the technology team – the very people who will be in charge of the most expensive transformation that underpins the deal.

As we've discussed previously, the risk of losing employees from the joint team looms throughout the project delivery. The loss of employees in this context could mean staff actually leaving the business, or losing interest but continuing to work in the business. These actively disengaged (yes, that's a real HR term) staff are very difficult to manage and they work hard to erode project confidence throughout the delivery. They won't be shy about letting everyone within the team and the wider business know how they feel about the changes ahead, and often paint a doom and gloom picture.

To tackle this, the Engage process requires:

- Target company research
- Conative profiling

- Tech profiles
- Team communication

I appreciate that these additional tasks may seem unnecessary, but for relatively low effort, you can identify potential problems early and get staff on your side to deliver a more effective transformation.

Examine. IT due diligence isn't exciting. At its core, it is an auditing task. A clichéd phrase used in this context is, 'You wouldn't start a long journey without a map', but if you expand the saying to, 'You wouldn't lead an expedition involving thousands of people without map', I think it's more fitting. If the team is missing even 5% of the information it needs, you may have to go cap in hand to the board for extra time and resources. Simultaneously, you can't spend too much time analysing systems.

Because of its scope and complexity, IT due diligence within PMI presents significant challenges. You have to deal with:

- Multi-party/multi-company technology teams
- Differing cultures and technology policies
- How technology is implemented in each firm
- The entry requirements across firms – what is acceptable at one may not be at another
- Software potentially being moved to another company
- The financial impact of transferring or novating software licensing

- Delays caused by technical unknowns or vendor support and flexibility

In addition, it is critical to understand if what you discover is acceptable in the new world technology environment. For example, the acquired business may be running a critical application on a configuration or platform the new firm deems to be unacceptable. It is almost certain that a high-level view of systems will already be understood by the purchaser, as this will form part of the M&A deal, but this may only cover a percentage of the application estate.

While a company may have a small number of 'above the line' systems, they are likely to be running thousands of applications. These all need to be assessed and must adhere to the plan.

During the Examine phase of the Managed M&A, your team will need to:

- Interview IT stakeholders to gather key information
- Develop a detailed asset inventory/configuration management database (CMDB)
- Develop a project-specific catalogue of services that can be used to plan and manage the transition
- Create and manage the list of support contracts
- Create a company-wide software inventory
- Ensure the licensing implications are understood
- Record IT finances and contracts

Regardless of the effort you expend, there will inevitably be some surprises on the way, but it is imperative you carry out a thorough and quality assessment with the limited time and resources that typify PMI projects.

Envision. Everyone's eyes will be on the transformation team's proposed plan as they will want to assess its realism and associated risk. I'd argue that a good project plan is in fact a sales proposal that will be circulated across the board. It needs to be convincing and it needs to work, demonstrating:

- You understand the companies' business
- You have key production services in hand
- Data and services will remain intact
- You've gathered engineering input and included it in the plan
- You've highlighted any R&D activities and potential costs

During the Envision phase, the technology architects representing the different parties will establish whether the assumed synergies between systems will in fact work and develop the associated technical design. They will solution-test end user applications, and the phase will complete with a detailed dress rehearsal to test the solution.

Summary

When conducting initial workshops or even sales meetings with clients, I will often give them a summary of the Managed M&A

methodology and the benefits of taking a 'people first' approach. Frequently, I see nodding and gestures, but the client wants to get straight to the point – to the integration, where the action 'happens'. Often this is because the assessment of companies and teams is initially seen as a waste of valuable time, but this soon changes when the meetings between the joint project teams become overly long and difficult.

This chapter therefore gave a brief overview of what is thus the essential 'people first' approach to PMI technology projects –

- Firstly, we outlined some of the key questions that you will need to consider if you want to be well prepared and achieve the best possible outcome for your PMI project.
- Secondly, we explored the Managed M&A method – key to which is spending a small amount of time assessing and 'architecting' your team.
- Lastly, we dug deeper into the key foundations of the Managed M&A – Engage, Examine and Envision.

Now we have described the three phases of the Managed M&A, we can explore each one in detail.

Chapter 6

Engage

The function of leadership is to produce more leaders, not more followers.

RALPH NADER

My origins and extended family are from St. Lucia in the Caribbean. During one visit, a distant relative suggested he take me on a catamaran tour. I jumped at the opportunity and was pretty excited, until I spoke more to my relative. It transpired he had a university degree in computing, but the only work he could find was serving rum punch to tourists.

On another trip to Northern India, I took some technology classes to see what they were like. I'd been in the industry for years, but I'm always curious to see how people work. I wasn't prepared for the long hours – we were working from 6am until 9pm with no breaks, followed by some beers in the evening. It was an exhausting experience, but the locals took it in their stride. They really wanted to learn.

I often think of these two scenarios. People in these countries sacrifice so much to be given hands-on technology experience in a business. Compare them to technology staff in Western businesses who often have access to an abundance of tech that they don't maximise. I've certainly had periods like that in my own career, when I didn't take advantage of what was sitting in front of me: not learning the latest and greatest tech, not really progressing, and not engaged.

Engagement is a key concern, so we'll be spending quite some time on it.

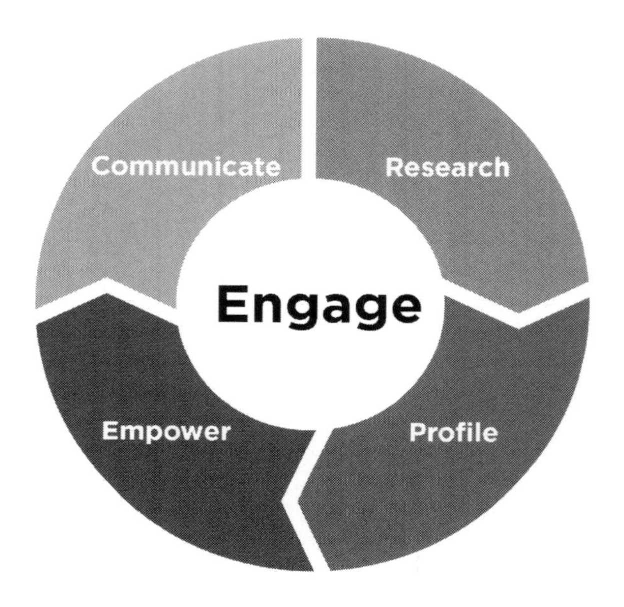

According to Gallop, only 13% of employees worldwide are engaged at work.[1] That equates to just one hour of engagement at work a day – at best. Introduce a major event such as an M&A deal, and the risk of impacting engagement increases significantly.

The main challenge during the project inception is motivating staff who have been hit by sudden uncertainty. 'People are your best

1 http://news.gallup.com/poll/165269/worldwide-employees-engaged-work.aspx

asset' is an overused phrase in business, but in an M&A PMI programme, they are your *only* asset. Therefore, engaging staff is the most critical requirement of a PMI project, but it's one that is often mishandled, and the impact of this mishandling is not realised until people either leave the company or there's a failure.

Dealing with career concerns during these transitions is assumed to be a Human Resources function. Indeed, it is for the general 'What's going to happen to me?' type of questions, but HR is unlikely to address questions specific to technology teams, such as new IT working practices and equipment. To be fair, HR staff don't have this kind of specialist knowledge, and they do have the wider team to be concerned about.

A good approach is to look at the project from the tech team's perspective and identify how the PMI could be an opportunity for them to up their game and shine. If they are only engaged 13% of the time, they have plenty of capacity for personal career development. Unfortunately, some people cannot be helped – they will simply sleepwalk to redundancy, no matter what you do. But others will express a desire to learn and grow.

For example:

- A talented software developer might be deeply concerned that they could end up administering and maintaining code rather than creating it within the new organisation
- An engineer could learn and master a new technology
- An applications architect might spot an opportunity to work on larger-scale enterprise applications

- A server engineer might want to move into a business application role
- A server team manager might be keen to lead a bigger team and maintain a hands-off position

You have a good chance of preventing staff attrition if you take the time to understand the team members and their skills early on in the process – it is possible to find a win-win solution where they can learn more skills or grow professionally, and you get to keep them on the team. Focus on the development of the technology staff at the core of the delivery. It will help address a primary challenge and deliver a smooth(er) integration. It can also be very rewarding to watch people grow through adverse times. By conative profiling the staff to identify their innate skills and assess their technology skills, you'll be able to see if people are in the right jobs – both from your PMI perspective and in the context of going into the new organisation.

Assumptions quickly become apparent when a PMI commences and the deal is announced to the technology employees. These assumptions will depend on whether this is the first time the company has been acquired or not. Quite naturally, they are often based on fear.

The two most common assumptions on each side are as follows:

Firstly, technology employees from the acquired business will assume they will be made redundant or replaced soon after the acquisition. If they do not lose their job, other changes may involve moving offices or being 'TUPEd' into a services firm, which will impact their lifestyle. With a negative mind-set, their imagined fear can become a self-fulfilling prophecy.

Secondly, with limited information, time and resources, it is likely that the buy-side staff will make assumptions about the capability of the target technology team and the value of the IP their company has purchased.

NOTES FROM THE FIELD

Synergy assumption

Back in 2005, eBay acquired Skype, the online telephone business, for $3.1bn. eBay assumed that if it integrated Skype into its platform, buyers and sellers could - and would - talk in real-time, creating a better sense of community across the globe.

At the time, eBay was the leading online retailer worldwide. Both companies had a huge global footprint, so you can imagine the PMI effort the acquisition took to deliver. The problem was that eBay hadn't correctly assessed its user behaviour. Sellers were simply too busy; they didn't want the interruption of buyers calling them, potentially 24/7, so they rejected the new Skype features. Realising its failure, eBay sold Skype for £2.75bn four years later.

Comparatively, this doesn't seem like much of a loss. But consider the integration and management efforts which would have stopped eBay from developing and improving its existing core business activities. Over this same four year period, eBay's core business reduced, buyers and sellers started to leave the platform, and Amazon took over as the leading online marketplace.

At a fraction of the cost, eBay could have trialled a mock-up of the proposed scenario worldwide without actually purchasing Skype. If it had engaged a small number of eBay buyers and sellers, the company would have realised its mistake without eroding the value of the business and spending billions.

The lesson: if the initial assumptions are incorrect, this mistake can come at a great cost.

Using the same company, here's a contrasting example:

NOTES FROM THE FIELD

Synergy heaven

The eBay purchase of PayPal back in 2002 is a perfect example of cascading customer synergies.

Prior to the purchase, eBay was transacting without a primary payment mechanism. This impacted the bottom line, because if an eBay buyer and seller could not agree on a payment method then they would not proceed, which meant less commission revenue for eBay.

When eBay purchased PayPal, it changed its software to entice PayPal customers to come on to the eBay platform. This cross-promotion led to an increased volume of people buying from each other, growing to trust each other and the service.

PayPal had been doing perfectly well prior to the acquisition, but through proper integration with eBay, it became the most prominent online payment platform. This in turn led to growth.

As the synergy developed, so did the united platform. For example, when a new user created an eBay account, a PayPal account was also created for them. Usage of the PayPal platform increased due to the new eBay users, which provided PayPal with plenty of evidence that its finance platform was vast and trusted. Consequently, PayPal could persuade other ecommerce companies to come on board too. As a result, PayPal became even wealthier independently of eBay transactions.

Soon, people were sending money across the world via PayPal, which created more profitable revenues than national transactions. The synergy between the two companies was significant.

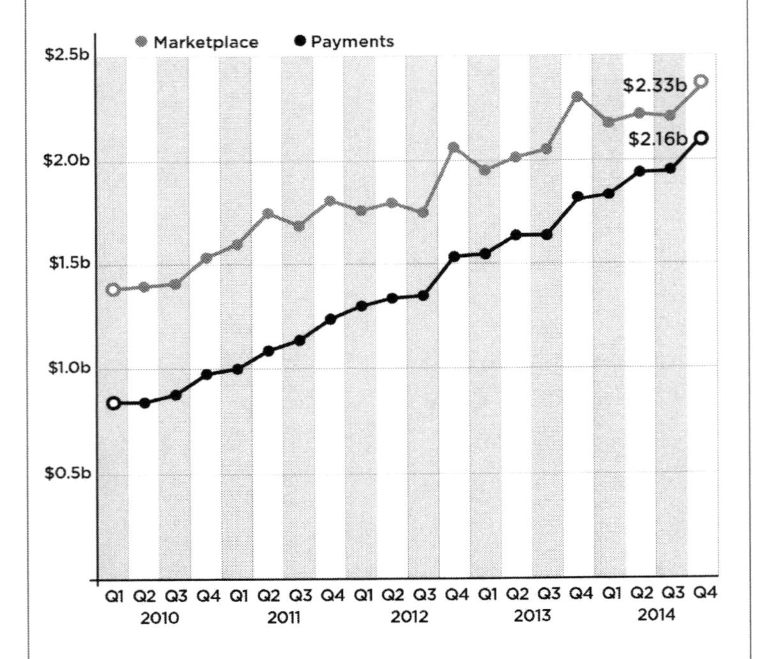

eBay bought PayPal for $1.5bn in 2002 and created a separate spin-off PayPal business in 2015, which was valued at $40bn at the time. PayPal's market capitalisation ($80.25b) has grown to nearly twice that of eBay ($39.21b).

The lesson: some pay-off!

During a PMI, the tech teams from each business are likely to have different cultures and will need to adapt, especially if they end up co-locating to work together in the same office(s). There are clear benefits to cross-pollination, learning and retaining what is best from both businesses, as highlighted by the successful eBay acquisition described above. If the technology employees from the buy-side are prepared to make changes to both processes and culture, they are more likely to achieve greater rewards from the synergy. If teams are being merged, you need to assess your own/internal team too.

So let's get stuck in with seven activities to help transform the technology staff's engagement.

Empowering staff

As soon as the deal is announced, the staff are thrown into a state of confusion regarding the future of their company, their own professional position, their current projects, and the impact on their personal lives. A technology employee may keep their position, but the technology culture and practices could be totally different in the new firm, and this may make them vulnerable if they do not know what to do. Soon after the announcement, technology staff are often requested to work on the necessary IT transformation, but they have not been consulted on how the changes will affect them from a technology perspective.

Due to the need for the multi-party team to work rapidly through the typical dynamics and activities that occur during the formation of a team, going above and beyond their normal duties, it is likely, as with every transformational journey, that things will get worse before they get better. Bruce Tuckman, an American psychologist who worked for Princeton University, is best known for a short article named 'Developmental sequence in small groups' in the 1960s. This article explains how teams form and work together, and coined the phrase 'forming–storming–norming–performing' to summarise common challenges and milestones teams reach before they become fully operational. These four milestones often happen in a very short time frame.

With Tuckman's model in mind, you can prepare your teams to:

- Consider and plan for the four stages of the model – what can you do to make each stage more effective?
- Set clear expectations for the team, for example, make them aware of the expected journey
- Make the team aware of the upcoming turbulence, and of the plans to smooth things out
- Ensure the team understands that there is opportunity in every project to grow and develop personally, if they are willing to participate

If you lose key personnel, you may have to bring in contractors to replace them who will not have the same in-house experience or knowledge. If you have to hire new staff, there is also a hefty cost attached. According to Louise Mallam, a leadership coach in the UK and founder of the company Leading Edge Performance, when it comes to staff replacements, you need to factor in 120% of their salary to cover the costs of recruitment, interviewing, marketing and the twenty-eight weeks it takes for new staff to become effective. So a non-leader on £25k a year costs £30,614 to replace; a high-level technical person on £60k will cost you approximately £72,000. For leaders, the true cost can be up to 250% of their salary. When you look at these costs, it is clear there's a benefit in investing time and effort into profiling your team(s).

Ensure you answer some basic questions to help alleviate their concerns:

- Why was this M&A deal made?
- How does it benefit the firm?
- How will customers of each firm benefit?
- How does the mission of the acquisition align with the mission of the buy-side company (also known as the strategic fit)?
- What type of investment is this – is it intended to reduce costs, increase market opportunities, or was it a competitive purchase?
- What assumptions have been made regarding the joint opportunities, also known as synergy opportunities, that can be gained from this deal?
- How does the deal benefit technology employees?
- What will the new team look like?
- How different will the new environment be?
- How will IT access change after this project?
- What opportunities will staff have to grow?
- Where will staff be working, and with whom?

This will create a sense of openness and positivity. But in the early stages, not all of the answers will be clear, so how do you distribute the answers to the staff in an efficient way? How do you highlight the differences between the two companies and emphasise the inherent opportunity in the PMI?

You create a brochure.

A brochure sounds old fashioned, especially in print, but big firms like Compaq (acquired by HP) and McDonalds both built empires from a brochure. Brochures help you articulate clearly what you are doing and help others understand your vision. A project engagement brochure will prepare your joint technology team and help the transition.

The brochure needs to:

- Provide brief information about the companies and the technology teams
- Set out the expectations of the PMI project, predict turbulence, and explain what is expected from the teams
- Outline the potential opportunity to learn
- Answer the common strategic fit and synergy questions
- Explain the potential benefits of the project
- Outline an integration timeline if possible
- Introduce service management staff from the buy-side
- Highlight current technology practices and potential differences
- Explain the security differences

Collating the information required for the project engagement brochure will force you to uncover and document the foundations of your project, but not from a traditional project initiation perspective. The brochure will help the joint project team by providing a clear and concise introduction to all the parties involved and an opportunity to be more aware of what's happening. It has the added benefit of demonstrating that your PMI team is experienced and well-structured. I would personally opt for a printed brochure over an online or intranet one as staff may not look at online content when it is first published. A printed document will sit on their desks, and they are more likely to flick through it during downtime. I also recommend a video brochure, but only if you have the skills to create something convincing and highly polished.

A template for a project engagement brochure can be downloaded here: http://info.beyond-ma.com/brochure

Company research

I highly recommend you conduct independent company research before meeting technology staff from the target company. Research provides you with a better background to the organisation, and you'll be able to demonstrate genuine interest in it. Note the operative word here: independent. Don't rely on the material or information from the M&A deal, which while useful, will be focused on the acquisition, not the integration. If you don't have an independent view, how can you have an independent approach? Doing your own research will allow you to avoid the rumour-mill that may be churning within both the buy-side firm and the target firm. In addition, you will gain a lot more information from speaking with the staff across the hierarchy.

Below is a table showing potential sources of information, the methods that can be used to collect it, and the type of information and flags to look for:

Location	Information	What to look out for
Endole/DueDil (Both commercial websites)	Enhanced search of Companies House	Basic company information Company structure
National and local press Google News search	Investigate how the target company is perceived by the public	Unless an article has made it to the first few pages of a national newspaper, it is reasonably safe to assume that the readership of the article will be low. This information gives an instant view of the company, the spokesperson, and the culture. Going back over the past year, you may be able to spot some issues with the IT systems that have made it into the public eye. For example, a customer our team worked with was at risk of being taken off a secure Government network due to non-compliance of their IT systems. With another customer, a local newspaper was querying the benefits of a recent merger as the predicted cost savings and efficiencies had not been met.
Corporate website	Information on the company's organisation and culture	Tone of the website copy Values 'About' page

Location	Information	What to look out for
Glassdoor https://www.glassdoor.co.uk/index.htm	How past and current employees consider the workplace	Look out for reviews from the tech team
Wayback Machine Browse http://archive.org/web/ and provide the company website URL	Determine how many changes have been made to the website	This site provides some useful insight into previous strategy: how long has the site been operational? Has the message, and therefore strategy, changed over time?
Annual report From the corporate website	Annual reports often contain key information that can provide some additional clarity about the company and its technology	They take a long time to read, so you might want to find a good researcher
Share price http://finance.google.com	Identify challenging periods	While many of the IT team will not be aware of the company's financial performance, it may reveal periods of time when working there was either easy or challenging for them.
Twitter Search the Twitter handle	Social interaction	Customer feedback/complaints
LinkedIn	Social interaction	If you have a LinkedIn Premium account, you can see the company's total employee count over the past two years, how actively they post, etc.

Location	Information	What to look out for
Facebook	Social interaction	How active is the page? How much interaction is there? What is the number of followers, likes and complaints?
YouTube	Social interaction	YouTube channels are usually rich with marketing and case study material.

The company's digital footprint and social media presence can provide a gold mine of information, but it takes time to research. Fortunately, tools such as Fanpage Karma exist. Fanpage Karma provides valuable insights on posting strategies and the performance of social media profiles, including Facebook, Twitter, and YouTube.

Office/working environment assessment

As a consultant, being allowed to enter a customer's workplace is a privilege. The initial visit after the acquisition has been made public can be like entering a war zone, but the location of the office can tell you a lot. Is it inner-city, the suburbs, or the countryside? Pay attention to the office itself, the exterior, reception, and communal areas.

If the technology staff are moving office location as part of the project, the difference between where they work now and the future office can impact future engagement. You're not going to be able to change any plans to relocate staff, but you can help their transition.

When you visit the businesses, you will be privy to certain signals that will give you clues as to how the team operates. For example, as you walk between employees' desks, it may be possible to identify pockets of information warriors. These are people who self-teach and read the books that are on their desks. Even in this world of digital books, technology books tend to be weighty physical tomes. If the books are gathering dust, explore why that is. Have the staff been experiencing problems learning new skills?

It's difficult to predict exactly what you'll find, so I'll share some of my experiences. One company I worked with had modern, trendy offices in London, which seemed vibrant and busy. However, the annual report revealed that an accident had occurred two years previously that had damaged the brand and profits. The tech team were busy trying to help a business struggling to recover.

At another company, I received an extremely friendly welcome as soon as I walked through the door. There were smiles all round, an impressive board room, but the tech team was neither friendly, nor pragmatic. They were married to the ITIL book of best practices, which translated into every change becoming difficult to deliver.

As you will no doubt have experienced yourself, things usually turn out differently to the initial observations you make of the project and team. So why not make it a conscious activity to assess the environment and people, rather than leaving it to chance? A guide to help you record your observations can be found at http://info.beyond-ma.com/CompanyAssessment

Technology team profiling

Conation is a word that dates back to Plato, yet we seem to have lost the use of it.

> [Conation is] *The mental faculty of purpose, desire, or will to perform an action; volition.*
>
> Oxford English Dictionary

I found this word difficult to grasp because it asks you to separate your natural behaviour from your personality. I thought they were intertwined, but there are three parts of the mind:

- Cognitive: your thoughts
- Affective (or personality): your feelings
- Conative: your actions

Conation is a key objective for finding out about employees' strengths by assessing their natural instincts. If they are stressed, you can identify if the M&A is the cause or whether they have been working in the wrong job.

This is at the core of engagement. It is standard to assess people's skills as you start to work with them, but I am also recommending that you run conative tests to understand how each individual approaches work most naturally. Assess the team as a whole and make slight changes as appropriate to get the best performance from them.

Yes, conative profiling is an unusual approach. But the PMI itself is an unusual situation for most people involved, so it is entirely reasonable to approach the project and the team in an unusual manner. Profiling employees will probably save you a couple of months, as well as reduce stress and sleepless nights. If you do it correctly, you'll understand the make-up and dynamics of the team within a matter of days, depending on the size, location(s) and attitude of the team. Otherwise, you won't get this understanding until you've worked with them.

To profile your staff, establish the following:

- Who are the staff?
- What are their technology skillsets?
- What are their innate skills?
- What do they want to learn, if anything?
- Why do they want to learn?
- What work do they enjoy?
- What is their approach to work and communication?

It may seem like a lot, but it's pretty simple.

Professor Carol Dweck is a professor of psychology who teaches courses in personality, social development and motivation at Stanford University. Dweck is famous for identifying the concept of people with a growth mind-set, which she demonstrates against a fixed mind-set. You are more likely to encounter people with a fixed mind-set – people who may appear to be stuck, negative or opposed to change. This is where the conative profiling really works because you uncover and identify everyone's strengths. By embracing their potential, you stand to create a more effective and happier team.

Technology staff who have been working in the same production line position for many years may come across as closed-off or negative, but they may once have had a passion for technology.

They may actually have a growth mind-set, but it's buried deep down. The PMI project is a good opportunity for them to get back on the horse.

In the past five years, I have noticed more aggressive engineers across clients in different industries. It is common to learn that the aggressive technology people are in fact some of the best delivery personnel. The trick is to get to know these people, taking a different approach to engage them. The conative tests will work here too, and as they involve so much detail, highly skilled techies will love them.

People profiling

Profile assessment tools are at the heart of the Managed M&A as they are used to highlight the behavioural differences between employees. Participants are asked to answer a series of questions which provide a detailed report about their personality and behaviour. You can learn how to interpret the tools yourself or engage a qualified consultant (with a technology background) to run the assessments and make recommendations for positive change.

Note: this section recommends using psychometric tools. Their usage will need to be cleared with the target company's human resources department.

There are many different profiling tools – most are 'affective' assessments, in that they test your personality on the basis of your preferences and desires. I used to use DISC, which is a widely-used personality assessment tool that helps identify personality traits, but now I use Kolbe, a conative test that helps identify how people work most naturally, and what constitutes their real selves. What I like about Kolbe is that it does not ask people to change, but rather to be obstinate in being themselves.

Kolbe allows you to go deeper to understand people's natural strengths and you can use this knowledge to align them to the type of work they will perform best, so I find it more suited for team transformation. For instance, up until recently I used write software (and I'm quite good at it) but looking at my Kolbe Index results, I am a 'quick start', or someone who initiates work by taking a lot of risk, so I am more suited to the CEO/sales position I find myself in today. I enjoyed software a lot, but I love my role now.

Note that tools like DISC and Kolbe complement each other, but it's unlikely you'll run both during an M&A PMI.

Kolbe Index.[2] Kolbe Index highlights people's unique strengths. You can use the assessment results to understand how each individual prefers to work in the workplace, if that person is experiencing stress against their own expectations, and if that person is experiencing stress against their boss's expectations.

Kolbe Index assesses the following strengths:

- Fact finder – how an individual gathers and shares information
- Follow through – how they organise their work
- Quick start – how they deal with risk
- Implementer – how they deal with pace and tangibles

---

2 The Kolbe A™ Index and Action Modes® are the trademarks of Kathy Kolbe and Kolbe Corp. All rights reserved. Used with permission.

The Kolbe Index provides you with a unique index result based on four numbers, which is referred to as your MO (or Modus Operandi).

In the example below, the MO is '8-6-2-3' and if the environment permits the individual initiates any creative task or work by initially collecting facts before they commence. Their next action is to create a system or defined approach to said task:

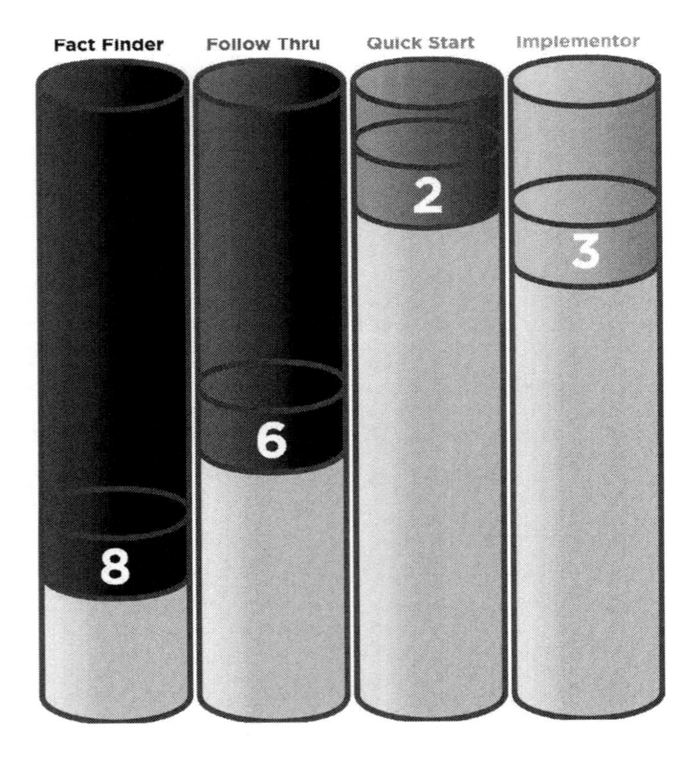

Note that in the diagram the higher number does not denote a better level of strength. A person who is a 2 in Fact Finder is not poor at finding facts, their strength is to take the facts and summarise them.

In this example, the Kolbe Index indicated he initiates tasks by finding out the facts first. The index also shows he also has a higher requirement of Follow Thru, so once he has the correct number of facts, he likes to create systems. If I ask him to put together a sales proposal without providing sufficient information, it will create (conative) stress because what I'm asking him to do goes against the natural way he prefers to work. Things get even worse if he is not allowed to develop the approach to the document.

As another example, I suspected a high performer was having issues outside of work. When she took the test, the Kolbe Index result stated she was in 'transition' – an outside factor was troubling her. She was surprised that software could pick this up, but the result was the catalyst for a discussion. After working with her for a number of months and allowing her time off, we retained a great employee as she had addressed her issues.

So, Kolbe can show how a person works and communicates most effectively and if you share your Kolbe MO they can potentially have a better working relationship with you. But how does this help with a PMI project? Firstly, you can assess each individual's natural skills, but you can also compare their skills against their expectations of themselves and against the expectations of their boss. By comparing skills against these expectations it is possible to see who is stressed or possibly in the wrong job. Secondly, you can combine the entire Kolbe team information and assess the entire team culture.

Kolbe and teams. The Kolbe model can provide a graph of the entire team's conative strengths and each team needs a balanced mixture to be highly effective. For example, 'Quick Start' represents how much risk a person is willing to take, and it makes sense that you cannot have a team that solely consists of risk takers (unless you are assessing a team of skydivers). So, ideally the team should have a blend of people who intimate tasks by taking risks, with people who are low risk takers and others who will moderate between.

In the technology industry, we tend to have a lot of fact finder/ follow through people due to the nature of the work. But having too many people with the same profile is a risk for a business because they are conative clones and are unlikely to spot gaps that people with other contrasting initiating skills would spot. It is likely that the team you are acquiring will be full of conative clones, so your job is to spot the talent across the different companies and, rather than just putting them together and assuming they will work well, work out how to shape a better, high- performing team.

Filling the gaps within a team doesn't always mean hiring more people. You may be able to move people around to create a better performing team. When you have visibility of conative clones, you can immediately look at addressing them. For example, if you have a load of fact finders, ensure they are being led by a quick start, or you could separate cloned workers to prevent them agreeing about their approach or work.

You can run Kolbe Leadership Analytics reports which take in all of your Kolbe assessments to provide information on:

- Team productivity
- Team efficiency

- How many members are currently experiencing conative stress
- People who are at risk
- Who is currently experiencing stress?
- Productivity – a chart depicting productivity shows what's missing and where you have too much of a specific skill

The assessment also provides information on the team's conative culture, which shows where the team is today, how the team members think it should be, and how the leadership think the team should be. Within the team of my own consultancy, the individual Kolbe Indexes showed that we were polarised. Some of these differences were obvious and we'd already worked out how to work together, but running the Leadership Analytics showed that we could benefit from some mediators and increase the number of quick start initiators, which was helpful as we were recruiting at the time. Now the right people are on board, we have seen an increased level of energy and more effectiveness.

Kolbe has been running its Index for over forty years, so there is plethora of information available at www.kolbe.com/solutions/for-business/.

Or to run a basic team assessment see: https://surveys.kolbe.com/survey/Team-Collaboration-Survey-for-Leaders.

Or if you want to run an Index and have the results interpreted for you personally by my team go to: http://info.beyond-ma.com/beyourself

Technical profiles

Often, you'll need to create a technology-based skills matrix to understand the capabilities of the acquired team. These matrices tend to be last-minute and home-made. Or you could use something that's well established and is an industry standard for assessing tech skills.

SFIA (pronounced 'Sophia') has become the globally accepted common language for the skills and competencies required in the digital world. It is the UK Government-backed high-level IT standard and describes the typical roles in IT and the skills needed to fulfil them. SFIA was developed by The SFIA Foundation, a non-profit comprising of five corporate members:

- The Institution of Engineering and Technology – The IET
- The Chartered Institute for IT – BCS
- The IT Service Management Forum – itSMF UK
- The Tech Partnership
- The Institute for the Management of Information Systems – IMIS

The Foundation developed tools to help companies map out what skills they need before finding people who have those skills. By placing people in the right jobs, the companies will ensure the team will work more efficiently and be happier, which in itself will attract the best talent.

In the longer term, if you can demonstrate how well you plan, structure and operate your technology team(s) against the SFIA framework, you have more chance of securing the best people. You will be able to demonstrate that you are taking an active interest in your people. From a PMI perspective, you can use SFIA to assess where team members are today, and use that as a benchmark to set targets to aim for tomorrow.

I worked with a technical employee who stated that they were philosophical about the acquisition of their company and generally felt okay with it. But through profiling and one-to-one conversations, it transpired that the employee was actually angry about their current position within the infrastructure team, felt limited by the opportunities available, and wanted to move into the area of business analysis. Following this, sitting in meetings with management from the target technology team, I noticed they had pigeon-holed the employee and had created a glass ceiling in doing so. The buy-side found a more suitable position for the employee in the new environment.

On another occasion, I met with a project manager who was struggling with the additional workload he had been given as part of a divestment. He had seen consultants working at his firm and wanted to give it a go, and his Kolbe profile highlighted that planning was not his forte – he was working against the grain with both his own expectations and his boss's. He was given a more suitable position, reducing the risk of the project.

For many Generation X employees, their IT skills have evolved organically. Their tech skills have been maintained without any focus on soft skills. Therefore, they have a general idea of their skills, but nothing has been recorded. Recording their profile also allows others to work with them more effectively.

You can help them to record their skills formally by meeting with them, making sure you record their aspirations. This is a much more effective and inclusive way to create the skills matrix than a few managers sitting in a room estimating people's capabilities. If you want to enhance this, combine what you learn from the employees' Kolbe Index assessment with their tech skills to ensure what they will do matches their innate skills.

You can find out more information about SFIA profiling at http://info.beyond-ma.com/SFIA

Communications

Tensions will likely rise during the project, so anything that can help create a better bond between the teams early on is a worthwhile investment. Proactive communication with the joint team members across all parties will allow the project to operate more efficiently and openly. But before you delve into digital communication, it's worth noting that one of the most effective approaches is to arrange a social event where the staff can meet each other outside of the workplace (location and budget permitting).

I have worked on many projects which had the intention and budget to set up social events, only for them to be forgotten due to pressure or the need to get on. Each time this has happened, it has taken a long time for the joint team members to get to know, trust and respect each other. Organising a social event is a small investment with a big return in regard to commitment and team cohesion. The team may even become genuine friends.

Something else to consider is the use of video. Video can help you market your message and updates, and give the communications a more personal feel. As people can access videos on their mobile phones, this is a good way of grabbing their attention for thirty to ninety seconds. In that time, you can relay a lot more information than you would by posting it on a blog, for example.

If you host the video on a platform such as Wistia, you can study the statistics to see how many people have watched the video and for how long. This way, you can determine whether people are informed and taking note of your updates.

In the UK, Mi Elfverson runs The Vlog Academy to teach people how to use video for business. This is a really helpful resource, more information about which can be found at www.vlogacademy.com/video-production/.

Sharing sensitive information across the joint team. Major challenges in sharing information between teams are data security and the sensitivity of information related to the project in hand. Hence, a secure platform is essential for joint communications of sensitive information. If either side leaks information about the IT systems, for example, this can create a serious concern and probably delays in the project, eroding your credibility.

There are a number of established virtual data room (VDR) products available, allowing online storage of documents that can be shared securely between two or more companies. Not only do these products solve the issues outlined above, many have also been proven in M&A PMI projects so you can focus on the project and not the reliability and security of new software.

Some of the leading VDR products are listed at www.capterra.com/virtual-data-room-software/.

Secure communications. As someone who is likely to work across the various parties, you have to consider how your team will store data that is predominately buy-side or target-side, and whether it should be separated. While the end result of the project will be a unified solution, during the transition all the companies involved often need to be treated as separate entities until the businesses are formally operating as one.

Running a PMI project involves recording detailed, sensitive company-wide information that is usually stored in one location or data room. This information may be application and services architecture from the CMDB and/or financial information such as support and software contracts. In the wrong hands, it could be a hacker's delight.

In addition to the location in which business data is stored, consider how this information is used and transmitted between the different parties. For example, if it is possible to email documents or attachments from the data room, this dramatically increases the risk of information getting into the wrong hands.

A security breach during the transition – for example, a situation where sensitive employee information from the target firm, authored by the buy-side IT team, gets into the hands of employees, or sensitive technical designs are emailed outside the business – creates uncertainty in regard to how the involved firms run their IT environments. It also erodes confidence in the project, which in turn will impact employee engagement. Ensure you speak with the internal security teams early to give them a heads up.

You may decide to set up a blog/intranet page to communicate to the team, but only do this if you have dedicated resources to update and secure it. In practice, I have rarely seen these being updated in line with the project, even with dedicated resources allocated. So often, the information is out of date.

Connecting in the new world. It is important to demonstrate to the employees in the target technology team that the new IT team is organised, prepared and welcoming. In the not too distant past, it was common for an employee to start a job in a new corporate company and find themselves twiddling their thumbs, because various things, like their computer login and building passes, had not been setup ready for their arrival. In the worst cases, new employees could be in this position for days.

But things have changed. You may have noticed a habit that people have formed – when they start at a new company, they take a photograph of their desk and workstation setup and post it on LinkedIn. Whether you find this interesting or not, the new employee is demonstrating that their company and environment are prepared for their arrival, and that they care. Aim for that same feeling and response when the technology team moves into their new environment.

Human Resources will deal with the office logistics and desk space, but further consideration is needed if employees are moving into a new technology environment.

On-boarding pack. There are many things that you can prepare in advance to provide employees with an on-boarding pack. There are some examples below:

- Introduction videos
- Project engagement brochure
- Technology management introduction
- The opportunity to meet their counterparts
- Office/data centre tour
- Technology policy outlines (personal development, change control etc.)
- On-boarding procedure
- Early visibility of their new computer equipment/ software
- Technology roadmap

All of these assets are selling tools that get the staff used to their new environment. Many will exist already and just need some reformatting before being included in the on-boarding pack.

Summary

In the Engage phase, if you give time and attention to engaging your technology team and helping them prepare for the new world, they will support your PMI transformation. You will reduce the risk of the acquisition being drained of value by ensuring:

- Less talent leaves the company
- The project will be easier to deliver
- There will be less risk of failure

The key to this is individual and team assessments so that you can put together a more effective group. By employing Kolbe and SFIA profiles, you will understand the team in much greater depth, and can use this information to plan and manage an effective and cohesive workforce. You can also use the information to set targets that help your PMI project and match the employees' innate skills and aspirations. You can then help reignite the careers of the technology team members, making sure they develop and prosper during the project so that as the company learns and transforms, so do they.

A PMI project can be a good deal for the technology employees, but remember there will still be surprises ahead. To reduce the element of surprise for the entire PMI team, we move on to the Examine phase.

Chapter 7

Examine

It's kind of fun to do the impossible.

WALT DISNEY

When I was living in New York, I was lucky that my next-door neighbour from London – let's call him John – was also sent out on assignment. I then made friends with a work colleague – let's call him Richard – who was from South Africa. John was a tall, genteel public-school-educated management consultant; Richard was an extreme sports fanatic. We planned regular weekends away, searching online based on Richard's sage advice that we must include the word 'death' in the search terms. Unless someone had died trying an activity, then it wasn't exciting enough.

In terms of our approach (not personality), we were poles apart. John was specific and conservative – a fact finder. Richard was a total risk taker, and I am someone who loves to innovate and try things. These differences were a good indicator of our conative instincts, but I wasn't aware of that term then.

We took a trip to Puerto Rico for a bit pf partying. A postcard showed a beautiful setting – a wide natural lake which looked stunning, but was on the opposite corner of the island to where we were staying. Nonetheless, we left the capital the next day to find this nirvana.

The hire car had issues, the roads were treacherous, and we had problems navigating because none of us could agree. It was dark when we reached our destination, so we ended up renting a room and having a few beers.

When we woke the next morning, we were surprised to see that the area was in fact a dump which looked nothing like the postcard. An entire factory complex had been built on the beaches, spoiling the view, and the only fun we could have was in the swimming pool in an all-you-can-eat hotel – for a fee. We spent the rest of that day driving back to the 'hungover' capital where everyone was recovering from the previous night's party which we'd missed, and we almost missed our flight.

We spent most of our time in the car, not at parties. We'd had a vision of where we wanted to get to, but had spent no time gathering any information about it.

Challenges

Discovery. As part of either the buy-side technology team or a consultancy representing them, you are likely to be one of the first people the target technology team will meet after they hear the news about the acquisition. You may face an immediate challenge here as you will need to obtain critical and confidential information about the business and how it operates technology before you have gained the team's trust. They'll be busy sizing you up. You'll need to determine quickly who the key stakeholders are (easier said than done in a large business), and then you will need to demonstrate that you have done this before in order to gain their respect.

CMDB. In order to deliver a successful project, you need an accurate inventory. A configuration management database (CMDB) may help create this as it contains information about the services, the infrastructure, and the relationships between them within a company.

Without knowing what is already in a company, you cannot establish what to do with its systems and who you need to work with to ensure they are transferred correctly into the new environment. This is the same for all IT projects, but the difference with a PMI is that it is a company-wide concern. Thus the importance of getting this right is amplified.

What's more, if you cannot determine what is in the company, you will be unaware of certain risks. For example, will the systems from the target firm meet the acquiring company's security requirements? Is some remediation work needed? Are there legacy systems that are on their last legs? It is common for projects to come to a complete standstill due to disagreement that has occurred as a result of an inaccurate CMDB.

NOTES FROM THE FIELD

Dangerous assumptions and faulty tools

Sometimes the CMDB/inventory solution is not technical; instead, it is about meeting the right person and asking them the right question.

For example, when I began work on an acquisition of a small development house of 200 employees, the client assumed it would take a couple of months to integrate as it had forty to fifty servers. On the first day of assessment, it became obvious that the company in fact had over 3,000 servers to be integrated, which obviously had a major impact on timeline and costs. The lesson: speak to people as soon as possible.

In another situation, I was working for a public sector client. Software asset management tools had been in place for a couple of years, but the information did not seem correct, and major delays were arising due to a poor inventory. After some investigation, I discovered that the configuration of the asset management tool was incorrect, and it was not storing the entire asset inventory. The lesson: do not rely on software tools.

Even after you create the CMDB, you will still face challenges:

- There is no single software platform that collects or stores information
- It's not sufficient to collect information and place it into a CMDB – you need action and potentially remediation plans for everything you have discovered

- The more you discover, the more work there is to undertake
- The more you discover, the more you discover, so when do you draw the line?

Service catalogue. You need to know what the services are to understand how they will be transformed (migrated, upgraded or decommissioned) and how they will work in the new world in order to be able to hand them over to service management post-Day One. It is possible that some of the bespoke software within the service catalogue is key IP driving the acquisition in the first place.

The primary challenge here is creating a clearly defined list that all parties involved agree to. In addition, the service catalogue will have contractual ramifications early in the agreement and post-implementation, because it will indicate which services will be provided and maintained in the new world.

Software inventory. Compiling the list of software being used by a company entails going into even greater detail. Even for a small business, such a list may well run into the thousands. You need a software inventory to understand what is in use, where it is in use, the financial implications of using it, and whether it will be allowed into the new company.

Support contracts. The PMI team needs to understand the financial burden of the technology service fully, and a major cost component is the ongoing vendor contracts. Your team needs to establish which support contracts can be moved over or novated across to the buy-side firm. By doing so, you will effectively be lifting up the carpet on the financial agreements between the target company and its suppliers. You may awaken a dormant account with a salesperson at a supplier, who may see the transition as an opportunity to mandate an expensive upgrade or transfer fee.

You will also need to identify which support contracts will take time to remediate. For well-known major vendors, they will already have written policies available online and partners who can ease the process. But for specialist software written by small software development firms, you may need to enter time-consuming negotiations.

In addition, it's crucial that you de-risk the whole process of reviewing the support contracts by establishing as much information as you can as quickly as possible. This will enable you to benchmark the current position and establish the level of work involved.

Finance. You need to understand the running costs in more detail than you will have established during the M&A deal due diligence phase. In addition to the usual facilities/hardware/software costs, it is critical to unearth any hidden and/or temporary costs, especially if the target business develops software. Your team needs technology experience so that they can correlate behaviour with cost.

Look out for hidden future costs, too. For example, if there has been a major software release in the past twelve months, is there evidence of additional (and expensive) contract software developers on the team? The target technology team may be running quite inexpensively today, but it may take up a lot more cash and resources in the future if major software updates are planned in the coming year.

Now let's cover each topic in detail.

Discovery

By discovery, I mean learning about the organisation. This differs to the research period in the Engage phase as that is about collecting information yourself. Discovery is speaking to people and finding out more about the business, and it covers the 'Interview' parts of the Examine model.

Time may be lacking, but it is essential you interview and collect information from people who represent the customer, service management, software applications and infrastructure team. This will help you identify any gaps or inconsistencies, which often leads to you discovering problems earlier in the process. During these interviews, a set of standardised questions can help you to find out more about the technology environment. Not only will the answers to these questions give you a holistic view of the business, they will also provide the target technology team with some assurance that you will consider their technology environment as a whole, leaving no stone unturned in order to reduce the risk of failure.

The table below shows an example of a standard set of discovery questions that will help you develop your PMI programme.

Discovery question	Further detail/why?
What are the current service desk tickets/ three- and twelve-month stats?	This establishes what the PMI team has taken on, and how will this environment deals with making change. Downtime and instability are unlikely to have been discussed in the boardroom.
What key projects are planned?	This is a standard project management question to establish what may interrupt or stop your PMI project from progressing.

Discovery question	Further detail/why?
Are there any scheduled change freezes that may impact the PMI?	You'll get some idea on key times when staff will be busy (a build-up to a freeze means high activity and less availability to help on other projects).
What key technology product launches are planned in the next twelve/twenty-four months?	What is already planned that may impact your timeline? Will the PMI team lose staff or focus during product delivery?
Are test environments available for development and infrastructure changes?	This strongly indicates the current technology change culture.
What is the change control process, the typical time to implement, and (in your opinion) the effectiveness of the CAB?	This is a significant cultural indicator, and very important as it is likely you will be implementing a wave of change that the team has never experienced before in the tech environment. You will need the change manager on your side.
Are there any compliance constraints on company or customer data? Are current industry standards in place, e.g. ISO-270001?	There's no point identifying, designing and attempting to deliver change upfront only to hit compliance, security or regulation issues, so you need to identify and assess this risk upfront.
What major incidents have occurred over the past twelve months?	This is a cultural and management indicator. If such information exists and it's possible to obtain it, that alone will demonstrate the openness and willingness of the team to deal with issues.
Are there any known issues that you should be aware of?	Migrating or changing a system or service disrupts it, and you may uncover an inherent problem. Avoid the PMI project being seen as the cause of an existing issue.
What nagging concerns do you have about the technology environment or services?	This uncovers issues that may arise or have been a long-term concern.

A more comprehensive list of standard discovery questions can be found at http://info.beyond-ma.com/Discovery

Ask for help. You require so much information that you need to identify who can help you gain it accurately in the shortest time. In addition, you will need to get the low-down on what's really happening at the coal face and with the production services. Any insider information that can help prevent downtime will help you to sleep better at night.

Identifying that contact or ambassador within the target company and gaining their trust is solely down to you.

Non-IT view of IT. Make time to speak to employees outside of IT to understand their view of the IT service. You may gain valuable insight into how reliable the services are, and also if the IT department is aligned to the business's needs.

NOTES FROM THE FIELD

Non-IT perspective

I worked with an IT department in a global retailer that went above and beyond to deliver a new, expensive service. Out of curiosity, I spoke to shop floor workers in some of the stores, who explained that their line management prohibited the use of this system. The IT department was therefore wasting time and a large amount of money running a system that wasn't being used.

In another situation, I worked with a tier-one university. Driven by competition, it provided amazing, and expensive, internet bandwidth to the students. From my interviews with the technology team, I could see that a first-class service was being delivered, better than many corporates at the time.

Things appeared to be going well, but then I had a few lunches in the staff canteen, starting conversations with random people. From the staff's perspective, the IT service was poor. It turned out that the technology team was slow to respond, often taking days to fix issues.

I would never have obtained this key information from the technology team as they were too close to the problem, but it was critical as resources and budget were needed to bring the services up to an acceptable level.

The lesson: speak to staff from different departments to get a fully-rounded view of the overall service.

Random drive-by conversations such as these are hardly quantifiable, but they generate insights that balance out the information you gain from the technology team. In addition, you can quote this feedback to reinforce the observations and arguments you present at the end of the Examine phase.

Stakeholder map. Using the information you've gathered so far, you can create a stakeholder map of the staff, their relationships, and how they can help the PMI project. A stakeholder map differs from an organisation hierarchy as it's a pragmatic view of the people who will be useful to the project. It is possible to group people together to have a visual representation of which parties or individuals may, or may not, support you.

For each party, you will need to establish at a minimum:

- The key players
- Their influence
- What's in it for them
- How they can help advance the project
- What are their key interests and motivations?
- What are their fears?

Once you have this information, it is worth assessing why certain parties are not supportive of the PMI. Taking the time to create a stakeholder map helps you find avenues to create better support across the multiple parties.

For a stakeholder map template, see http://info.beyond-ma.com/stakeholder-map

Group workshops. Please, for everyone's sake, resist the temptation to hold too many workshops. People often feel that the entire PMI environment is so complex it necessitates an inordinate number of workshops involving all the key people from all teams. In my experience, this can be a waste of time for everyone.

I recently learnt from a slightly distressed friend who is a senior manager in a technology business that his company has been acquired. His boss called him, excited, to discuss the news and tell him that they needed all senior managers to travel far out of town (on a Friday) for the first workshop to meet the senior managers of the buy-side business. The workshop was scheduled to cover twelve hours.

What is the subtext of this announcement? 'Now we've merged, things will become more bureaucratic and we'll expect you to work long hours. And we won't worry about the impact on your weekend.' If you start the entire programme asking people to work more hours without additional incentive, then be prepared for them to lose interest.

There is one benefit of group-wide workshops, which is to watch people's behaviour to determine who will collaborate and who will become an obstacle. It's worth having a workshop or two, but resist the temptation to have too many. Many people won't participate, and there is no major benefit to all the staff being aware of all the decisions made along the way. If you plan too many meetings, decision-makers and more skilled technical people will often miss them, rendering them ineffective.

Aggregate information from the discovery. During the early stages, you are essentially collecting the best information possible so that you can make judgements and respond to issues and risks. It is likely that if you have requested answers from more than one person in the technology team, you may get different responses.

As part of the Managed M&A method, you need to interview the key stakeholders using the standard discovery questions and create a heat map of the areas that appear positive to them versus areas they have highlighted as an issue. Throughout this process, you will need to draw on your personal judgement and experience to make your own assessments.

When you have all the information you have requested, you'll be able to determine who will collaborate and make the extra effort to work well with the team.

Configuration management database

A configuration management database (CMDB) is necessary to size and plan the project. The art of CMDB creation is in presenting a realistic assessment of the IT assets while not over-complicating matters by providing so much information that you cannot make any changes. You need to get this right otherwise you'll lose your credibility.

A CMDB can be an opportunity to build a case for a project extension as the inventory may end up bigger than you anticipated, or you may need more time to transform business-critical systems for the new world.

Note: in a *People First* approach, the creation of a solid CMDB can be a worthwhile project to assign to one or more of the tech team to give them an opportunity to stretch or demonstrate their skills.

This can be a sensitive area, as no one wants to admit that they don't keep an up-to-date CMDB, and they'll assume that everyone else does. They are wrong in this assumption; many companies struggle to keep an automated register and only collect information when they're embarking on major transformation projects. Once the project is over, they may scrap the information they collected, or not update it.

The approach to creating a CMDB will depend on the systems, existing tools and information, and the technical staff's availability. The following tips will help:

- Assign someone to oversee and be accountable for the CMDB creation
- Ensure they have a fixed, enforceable deadline for the CMDB creation
- Ask them to send progress updates to your team to reduce surprises and delays
- Assign a talented technical person who can script or code and a business analyst to gather information together
- As a general rule, avoid manual collection of information
- Avoid the team working from information on spreadsheets, encouraging them to get hands-on access to assess systems and collect information. They will then get a feel for the technology, e.g. its load and reliability
- Ensure your team checks the retention settings on software-based asset management tools
- BAU staff may find gathering this information a challenge, due to other commitments

Not only do you need to understand what systems and assets are currently in use, you need to understand who uses them.

Hopefully, the in-house software tools will cover this, but there have been many occasions where customers have had to assign engineers to collect this information manually. This is a major expense that management is not usually aware of.

While collecting information manually is an expensive way to determine who is using IT assets, it does have one major benefit. You'll be able to speak to representatives across the business and get their perspective on the pending change. Then you can identify ambassadors who could be guinea pigs for future systems testing.

NOTES FROM THE FIELD

If you want to burn cash

A client was struggling with their PMI IT due diligence and was still assessing the technology environment after two years. The business had approximately 10,000 employees but only had a national footprint, so it didn't have the complications of multiple countries, time zones and cultures to slow things down.

The client realised the information from their expensive IT asset management system was inaccurate, so the company hired a number of IT contractors from a consultancy, put them up in a hotel, and asked them to call thousands of people via telephone to collect information and enter it in spreadsheets.

I'm sure you can guess what happened. By the time the project got moving, all the information the consultants had gathered was out of date, mainly due to corporate restructuring rather than tech changes. Due to the poor quality of the information, the client could not make progress.

The solution was to make one person in the internal PMI team responsible for the inventory. Creating ownership naturally caused that person to drive both the delivery and the quality of the results.

The lesson: the obvious problem here was the manual nature of the task, and I dread to think how much it cost overall. Be careful.

Some of the key areas you must ensure are assessed in the CMDB are:

Category	Description
Security	An entire book could be written about assessing security practices and solutions within a company. However, it is critical for the due diligence team to meet with security to understand basic principles and policies, gather corporate governance documentation, and understand any compliance frameworks currently in place, for example ISO or PCI DSS.
Software licensing	This must be top of mind and treated as an overall priority, as it's commonly an expensive issue. As an outsider, you are not privy to the local IT processes and procedures. If software is installed manually, there is room for significant human error, such as incorrect versions being installed, or unlicensed copies of niche software being used across the target company. Licensing is an important risk to assess in order to predict any future financial impact.
Networking	The network carries information in and out of the company. It may be used to transport data from the target company to the buy-side, and there may be a need for a temporary uplift in capacity. All of this needs identification, security clearance, planning, budgeting, and lead times.
External interfaces	No company operates in isolation, so a clear understanding of the information and data feeds in and out of the company is essential, depending on the migration method and the resulting Day One solution.

Category	Description
Custom code	Older, well-established companies will have code that has evolved over time to become business critical. These home-grown products may not need to be changed as part of the transition, but the buy-side needs visibility of what they are taking on. Some examples from my own experience: - An entire stockbroking platform written by a single IT contractor (unsurprisingly, he had his own yacht) - An entire finance system written with Access and Excel for a global firm - A bespoke customer record management system (CRM) built in-house within a corporate - An in-house-developed billing system that technical staff were too scared to change, and which the business was 100% dependent on No one really knows how these systems work, and no one wants to change them.
Systems management	The systems management/monitoring tools will provide a good insight on how IT is operated. Some questions to ask are: - Is anyone monitoring these systems, and are they kept up to date? - Are there any recurring issues? - How are critical/public-facing production systems benchmarked? This is important as the benchmarks need to be the same or better in the new world IT solution. If you don't record anything during the Examine phase, you won't have anything to refer back to - Do the monitoring tools operate on any custom code, and are the developers still on site? This may highlight hidden costs and extra risks.

Category	Description
Capacity planning	What capacity management plans are currently in place? This is usually monitored by systems management tools, but an overview is required in the high-level design (HLD) – is anything at max capacity now?
Performance	This is usually monitored by systems management tools, but an overview is required in the HLD. Is anything showing performance issues at the moment?
Data discovery	Take time to establish what data risks you may be facing –again, you cannot assess the risk until you know what's there. A common mistake is to assume that data will be picked up as part of assessing and moving applications and services, but it's a separate task. What are the data movement/compliance policies?
Data saved locally on devices	There may be a risk that critical information is stored somewhere it shouldn't be.
Public cloud services	These can be small department-specific products, such as MailChimp, or enterprise-wide products, such as Office 365. You'll probably need to record these services within the service catalogue, not the CMDB. Do not assume that public cloud services will be permitted by the buy-side IT security team. Get clearance, early.
Printing	This is often overlooked. You need to find out about how printers are maintained and how they may be used if equipment is moved.

More information on the CMDB can be found at http://info.beyond-ma.com/CMDB

Service catalogue

The agreed service catalogue will become the list that the team works from to understand what services they need to provide. It will likely determine some of the work streams on the project plan and the approach of the entire project.

The operative word here is 'agreed' – it takes time to identify the services and negotiate what will be integrated or transformed as part of the PMI.

If a current service catalogue is in place, using this as a basis to produce an agreed service catalogue as part of the PMI transformation can take time, and will certainly require resources. The agreed service catalogue will be utilised post-Day One for service introduction into the new environment, and it will become critical after the 90/180-Day post-implementation review. Hence, it is important to see it as a living, working document, not a disposable one for one-time project use. Ultimately, you need to provide a status of the services based on the M&A transformation programme.

Categorising the catalogue will help you to assess the variety of risks your PMI may encounter. To identify these risks, you will need to speak to the service SMEs to determine key information for each service in the catalogue in order to categorise them, establish ownership, and set availability targets. While you're having these conversations with the SMEs, it is likely their issues, concerns and any risks they envisage will become clear, depending on whether they are happy to disclose this information.

As an example, I spoke with a client who had been told, pre-M&A deal, that the software application estate they were acquiring was 100% web-based, and therefore transportable to different platforms. However, after categorising the services, my client discovered that although all customer-facing applications were web-based, this wasn't the case for the back-office systems. These back-office systems ran into the thousands, representing significant costs that my client had not considered. This is pure gold for consultancies, but something you as the leader of the PMI will want to avoid.

Software inventory

This is simply the list of applications (not services) running in the environment. A company may be running a couple of hundred services across its environment, but have thousands of applications to install. These applications will be running mostly on PCs and servers, but you may encounter legacy items such as old APIs and interfaces, mainframe or 'mini' computers along the way.

As with standard enterprise transformation projects, you need to whittle down and categorise all of these applications to critical applications only in order to make the PMI transformation manageable. This list will become your software inventory.

If the company packages its software, great. It may be possible to focus on the list of packaged applications and work upwards, instead of starting with all the applications and whittling them down. The people who packaged the applications will be a useful resource as they are generally able to script and automate, in addition to having a good knowledge of the IT landscape and the applications deployed within it.

The software inventory will, at the very minimum, list software, packages and comments/information, and identify which departments use the packages and how many they use. A template software inventory is provided online at http://info.beyond-ma.com/software-inventory

Support contracts

Make sure technology support contracts don't become a reason for delay. All PMI projects will require some time to re-assess the current vendor contract commitments, and this is an area for your procurement specialists and potentially your legal department, too, but you will need to ensure someone in the team has this in hand. All contract renewals, novations, and exits must be completed before the Day One cutover.

From a high-level perspective, you will need to know the impact of the contracts. The obvious factor is the ongoing costs, but you also need to identify any contracts (or lack thereof) for legacy software. It is worth finding out about any ongoing disputes with suppliers as these will give an indication as to how the company treats its suppliers.

It is essential to identify and single out any niche companies or services that may not have existing processes and documented policies on contract novation. Usually, niche contracts will take more time to negotiate versus the major vendors, such as Microsoft or IBM, because smaller software companies won't deal with transfers so often. There is thus a risk that they will carry hidden costs, which you will need to raise with the project board as soon as possible.

Your team will need to identify and categorise the support contracts into IT services, public cloud services, software, and hardware. While each type of contract will involve different key information, the main things to consider are the service level agreement, escalation processes, and costs.

After you've confirmed the contract details currently held on file, it is worth collecting the contact information and speaking to a vendor's account manager to determine:

- Any restrictions (e.g., if you're moving to new platforms or data centres)
- The cost of novation/movement into a new firm
- Any known lead times that may need to be factored in
- A review of marketing material
- A review of product details
- A review of vendor policies, procedures and quality management systems
- The vendor's ability to meet the needs of the project or department
- The experience and qualifications of the vendor's staff
- The company history and stability, including financial viability
- The vendor's capacity to deliver within the required time frames
- The aftersales service, including training
- Costs
- Whether the vendor is on the preferred provider list
- A summary of any recent inspections or audits

The PMI is a good opportunity to consolidate contracts and make better use of supplier agreements. A template support contract is provided online at http://info.beyond-ma.com/support-contracts

Finance

The PMI team will be making significant enterprise-wide changes, so you need to ensure you know the target company's current costs, even if these will change as the PMI transformation progresses.

Understanding costs is essential when you're considering a decommission or consolidation of services. If you can obtain the target company's previous two years' IT expenditure, you may discover some patterns of change in regard to staffing, the resource required to make change versus the number of staff currently in the target company, or even a change in technology that may have increased or decreased costs.

Detailing an exact approach to assess the technology budget could take an entire book, so for brevity's sake, I will refer you to a template finance budget assessment tool available online at http://info.beyond-ma.com/finance

Expenditure will be assessed by the wider acquisition team, but it is essential to assess the costs specifically from an IT perspective. Often, technology costs are hidden because either they appear in different cost centres, or technology is directly procured by employees purchasing cloud-based software-as-a-service packages.

Once you have collected the budget information, someone within the buy-side business will need to assess whether financial requirements will remain the same in the coming three to five years.

Summary

Throughout this phase, you will obtain a clearer picture of the target company's current environment, and it is likely the systems and services will be different to what everyone assumed. This is true for every technology project.

During the examine phase, you will have:

- Applied a structured approach and standardised questions to discover more detail about the environment
- Assessed or built a CMDB so you know what you are transforming
- Created and agreed a service catalogue to be used to track the project
- Created an agreed software inventory that identifies the business-critical apps
- Created a support contract to clarify costs
- Assessed technology finances

There are numerous factors that will dictate how long the Examine phase will take, but the most impactful will be the willingness and openness of the staff to engage and participate, and the deadline set by the project board for the Day One delivery.

If you've collected sufficient information and nothing has been hidden from you, you will now be in a good position to start creating a technology design blueprint and project plan, which is covered in the Envision phase.

Chapter 8

Envision

A committee is a group that keeps minutes and loses hours.

MILTON BERLE

In one position I had I was a technology trouble-shooter, and it involved so much travel that I used to take my passport to work – something would often crop up and I'd find myself on my way to the airport. I also was responsible for a team of fifteen people who were deploying our solution in various parts of the world.

We had been having problems with trying to implement a system in Brazil and every time we tried to speak to our representative – let's call him Ian – we lost contact with him. I found myself on a plane that day and in Rio many hours later.

Within minutes of arriving Ian greeted me, with a big smile – job done, I thought. As it was a long flight back I stayed for a few days. In that time, I learnt that Ian looked like a famous movie actor Brazilians loved. He now had a bunch of cool friends, a model girlfriend, and people offering him VIP treatment whenever he arrived at a restaurant or bar. Now I understood why the job wasn't progressing.

The point of this story is that while I was on this mission to deal with Ian, my 'anomaly', the wheels kept turning and the fourteen other people dotted around the world continued to deliver even though they were a virtual team. In my opinion, and from experience, it takes more than a plan to make this happen. Your team must really want to get the job done.

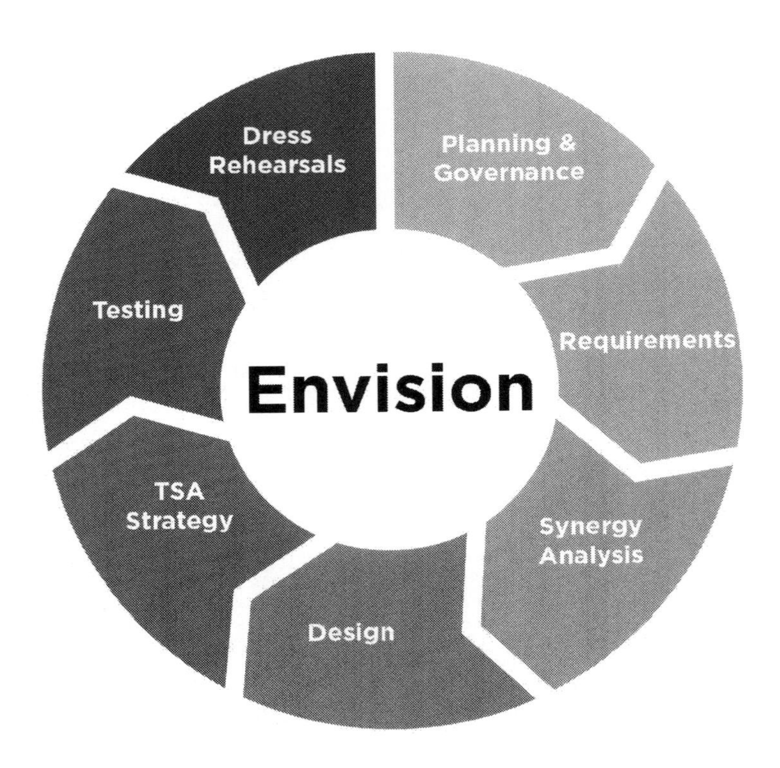

The technology solutions developed in this phase need to be aligned to the business requirements while obtaining buy-in from key staff from the joint PMI team. The intended outcomes of the Envision phase are a technology design and approach that deliver an effective Day One integration, and a realistic and achievable project plan.

Planning and governance. Your work will have a major impact on the employees and working environment of the merged/divested entity long after the technology project has completed. If you approach the project without an M&A-specific strategy, the team will pick up on this and stop responding. Over time, this dangerous situation will become apparent through either a lack of results, or many open activities that do not have a clear solution. This situation will escalate, because once the technology teams become dismissive or despondent regarding the project, they will tell their colleagues in the wider business. Then the Chinese whispers begin, and before long complete fabrications will appear.

If this is allowed to continue, the negative impact will far outlast the initial merger activity. The wider staff will view everything the technology team does in the new world as sub-standard. This may not be a concern for the M&A deal makers, but it creates a challenging working environment that increases the risk of the best technology employees from both sides leaving the firm.

One challenge that can arise relates to accountability, specifically who the project management office (PMO) team represents. Unless the team is built of resources from all parties, there may be a bias to their reporting – this is simply human nature. But if one side is dominant and running the project, the other side can become passive. Then, the governance will not be effective, as no one will be questioning the delivery.

Even with a PMO team in place, this does not mean that governance will be implemented effectively. Make sure the team doesn't simply take up seats in the board room.

Synergy analysis. The pressure of delivering assumed synergies can prevent the overall PMI from completing.

Assumed synergies are one of the foremost reasons for entering into an M&A deal, with synergies that support cost savings, rather than new revenues, typically being the more successful. Synergies that assume revenue increases from bundling or cross-selling into each involved business's marketing can be difficult to realise.

Realising synergies takes up management time, especially that of the technology management teams. It requires numerous outlays (e.g. major system upgrades and integration, and severance packages for certain employees), and sometimes further investment such as more integration of the IT systems.

Technical design. Other than the additional complexity, technical design is one of the few areas where there appears to be no difference between a traditional and an M&A project. However, you will be working between at least two architecture teams, who may have competing ideas and/or different approaches.

From a practical PMI delivery perspective, your technology design will produce the sales documents to pave the way to agreeing major change. If we treat technology design documents as sales documents, then we can approach them in a sales-oriented way.

Good sales documents:

- Are engaging
- Use original content, not boilerplate
- Are concise

- Sell the transformation, not the tools
- Demonstrate the value of the product
- Start with the end user journey/experience
- Provide clear guiding principles
- Align well to the customer's needs
- Use case studies as examples

It is essential to brief and engage staff in order for the teams to work together to deliver an agreed technical solution successfully. Without engagement:

- The teams will not work collaboratively
- There will be a major risk of not meeting business requirements
- The solution design will be excessively flawed (no design is perfect)
- Downtime is almost certain
- Financial impacts may be seen during the project and post-Day One

Therefore, aim to deliver a technology design that sells the change.

Transitional service agreements (TSA) strategy. TSAs are most often used in carve-outs where the buyer lacks the necessary technology capabilities or capacity to support the business on its own. TSAs are also necessary when the deal closes faster than the buyer's IT organisation can respond.

A TSA is a legal agreement, separate from the M&A purchase agreement, in which the buyer agrees to pay the seller for certain services to support the divested business for a pre-defined period of time. Understandably, the target business's CIO will not like TSAs as they effectively bind their team to services that they are no longer interested in, so expect a lengthy discussion and perhaps obstructions.

Common TSA challenges relate to the transition to the other company. For example, not putting all requirements and expectations in writing can lead to a poor definition of what will be managed and how these services will be managed. This poor definition of scope can further divide the teams and lead to misunderstandings, never-ending support agreements and rising costs. At the heart of these problems is a destructive mistake – not taking relationship changes and power-struggles between the businesses into consideration. Once the service belongs to someone else, it is natural for people to take their foot off the gas and provide a minimum service.

Dress rehearsals. The main objective of a dress rehearsal is to build more confidence in the plan and bring to light any assumptions the team may be making. Depending on the type of integration, you may not need a dress rehearsal to test the Day One cutover processes and assumptions.

While your focus will be on the technology team and the technical solutions, it is also important you ensure there is sufficient business engagement to test whatever you are changing. This may seem to be an obvious thing to highlight, but I have witnessed situations on many occasions where the business is surprised by or not aligned with the resulting systems.

You will need to establish clear accountability for any dress rehearsal activity – who will represent each party and ensure the correct resources are in place?

Planning and governance

Poor project planning in a PMI will have a major impact on the project manager and the team's morale. It is common to meet PMI project managers who are close to burn-out, overwhelmed by the sheer volume and complexity of the work. Most of all, they shoulder the burden of any project extensions or projects that are deemed failures. It is not uncommon for new project managers to take over the reins of a failing project with the benefit of being able to re-assess and re-plan it. But if the PMI project is thought through with sufficient time, care and attention, this problem can be avoided.

Below is a list of the common mistakes made by PMI project managers along with suggested remedies:

The wrong approach. It is common for a project team to approach a PMI project as if it were a standard IT project. While this will work to some extent, there is a risk of having to re-plan as they encounter unexpected problems.

Don't treat it as a normal programme. Get people involved. Determine your work streams. Make people responsible.

Not acknowledging the points in the project that involve pure guesswork. You may be planning an integration over many months or years, yet in reality, it is only possible to envisage and predict what will happen for a small proportion of the project timeline. After a few weeks or months (usually 90 days max), the plan becomes pure guesswork.

Be honest about the drop off points, and the times where the project activity can be guaranteed rather than predicted. Once you have established the areas of guaranteed versus predicted activity, focus the teams on the guaranteed work. This will help them work in small units and manageable phases without burdening them with too much of the project activity yet to be scheduled. Team members will then be fully focused on specific areas and can complete small units of work on time, often resulting in a happier team.

Focusing on the board. If the project manager is managing the delivery team and reporting to the project board simultaneously, they may encounter challenges as there will be a conflict of interest between these parties. While the joint steering committee/board is extremely important, if you focus on them and not the delivery team, the delivery team may not meet important deadlines, or may encounter major issues without you realising. Resource the project management team so that there is a project manager running the team(s), and another reporting to the board.

Starting off without a plan. A PMI plan can be created organically. With some hands-on activity, the plan may even turn out well. But starting without a plan and continuing in this way is likely to fail, mainly due to the multi-party nature of PMI projects. You require buy-in from all parties, who will likely want to see a plan on paper.

Not listening. The technology staff of both parties can provide vital information, but only if they are given airtime. Take time to engage with and listen to staff across the hierarchy, not just at the board level.

Too much detail. If a technical project manager creates the plan, there is a risk of it being difficult to decipher, follow, and maintain. Break the plans down into clear, simple steps, and strip out any unnecessary activity.

Too little detail. A common mistake is taking the high-level plan that was used to strike the M&A deal and using it as the project plan for the actual delivery. Because the original deal plan was at best a guesstimate, it will often have had little or zero input from the technology teams. The two plans need to be different documents as the real plan will change radically once you have consulted the team.

Another mistake is to use the assumed or projected Day One integration date as the drop-dead date that the team must work towards. This forms the basis of a rushed implementation plan.

Ideally, build an organic plan and compare it against what you sold to the board. You will need to explain the differences, but do not overly worry the staff about this until you are fully aware of the work that needs to be undertaken and what changes you need to make to deliver a solid integration.

Neglecting accountability. The Standard RACI Responsibility Matrix or other mechanisms will help you understand who is responsible for certain areas of delivery. This is project management best practice, but is often neglected. For a multiple-party PMI project, a RACI, or something similar, is a mandatory requirement, otherwise there's a risk that responsibilities will bounce between teams. Ideally, share, discuss and agree on the

RACI Matrix with your stakeholders before your project starts. Ensure every task has an individual accountable for it.

Neglecting official sign-off. It is critical to ensure that every phase is officially signed off before moving on to the next stage. Otherwise, there is a risk that something fundamental will need to be re-visited, meaning a probable project extension. Make sure you have a clear project charter and scope statement for all projects that have been developed with and signed off by the appropriate level contact.

If the programme/project is set up incorrectly, it can appear to be a huge list of tasks that will overwhelm the teams as soon as they look at it. You need to define the project work streams first, and then break these down into manageable tasks.

To see a list of potential work streams, take a look at www.beyond-ma.com/PeopleFirst/Workstreams.html

Accountability. You need to identify all the major project work streams and plan them individually with input from the SMEs across the joint project team. This is opposite to the traditional approach of developing a complex top-down project. You can distribute small work stream focused plans to the appropriate SMEs so they can develop the detail better than they would if they had the entire plan to assess. With support from others and better quality plans, you are less likely to burn out, and will provide a better oversight of the programme.

You need buy-in from the joint project team, and you need to be able to deliver the programme successfully. Therefore, I recommend that you assign each work stream to a primary individual in possession of suitable knowledge, but spread the responsibility across the team. This reduces the chance of a 'them and us' scenario and brings the team together, in turn reducing the risk of you shouldering the burden of the entire programme.

In addition, it is essential that you find the appropriate SME for each work stream, and ensure they are involved in the creation of the relevant plan. Senior SMEs may want to progress their career, so there is potential for them to take on an active role in managing their work stream, but I recommend that you have experienced project managers overseeing the work.

Planning hackathons. In the technology industry, hackathons are used to bring software, graphic and interface designers, project managers, and others together in order to run rapid software design and development with tangible results. Hackathons are seen as geeky endeavours, but don't let that put you off. They have proven to be extremely beneficial in PMI projects because they cut through politics and get results.

I've had tremendous success using hackathons within PMI projects and programmes. They take people out of their normal working environment and give them the opportunity to think clearly about how to get results. In addition, they are opportunities to ensure the joint PMI teams work together rather than in silos, creating more of a gel between the teams.

If you need concrete evidence of the power of hackathons and how they can turn an approach around, look at Satya Nadella, the current CEO of Microsoft. When he began his tenure in 2014, Microsoft was seen as a middle-aged IT company. Profits were down and it was competing against sexier companies like

Apple and Google, which were outsmarting Microsoft with better looking and more functional technology.

In fewer than three years, Nadella turned this perception around with his Cloud and Surface device strategy. He created large public events, reduced company meetings, and ran live hackathons to reshape the company culture into one that is now more innovative and collaborative. Part of his strategy was to re-ignite the creative processes in technical people and, in turn, transform perceptions of the business – which was helped by putting a few techies in a room to provide great PR and give rise to many new ideas.

How to set up a hackathon. From a corporate technology perspective, it is a simple case of ensuring you have the necessary resources (office space, computers, logins, software, and people), and then giving the people time.

The high-level approach our consultancy team takes is to:

- Ensure management agreed with the hackathon approach
- Create guidelines to ensure BAU work duties did not interrupt proceedings
- Find a dedicated space for the work to take place
- Create a small team – ideally three, but a maximum of four people with the correct skills and access who knew who to contact when input was required from the wider group
- Give the team clear directives of what the aim was
- Give the team three days (working nine to five) before they needed to report results

In one PMI project, we reduced a forecast R&D effort from six weeks to a few days. But the most rewarding aspect was seeing the small hackathon group beaming with pride at what they'd learnt and produced in such a short time.

In a successful hackathon, once the team sees the results, they will build momentum and cadence, which in turn ignites something really important. Employees who have been dormant at work for years suddenly start creating again. Instead of their working day being filled with process, procedures and form-filling, they are reconnecting with the reason why they became interested in computers in the first place.

If you have profiled employees as recommended in the Engage section, you'll be able to put together the perfect hackathon team based on the results.

There are risks that hackathons will not always produce the results you are looking for, but they are worth the risk because the team will get through an enormous amount of work in a short time.

There are some official guides to walk you through exactly how to set up a hackathon at https://hackathon.guide/

Synergy analysis

Most large businesses look at ways of growing via synergy. Some invest in meetings and international retreats to develop ideas, products, and markets. Effectively they are exploring what to acquire, and why. The M&A team from the buy-side firm of your PMI project will have clearly established the synergy opportunities that interest them well before you got involved.

You and the technology team need to know any assumed synergies and keep them in mind when you're making technical and action plans. The challenge can be obtaining this information in the first place, because it may be that only a few key people know the real reason for the M&A deal. But the actions you take today could impact synergy realisation in the future.

> *Most merging companies entering a deal don't have a clear understanding of the level of synergies they can expect through increased scale.*
>
> Bain, management consulting firm

The bottom line is that big businesses desire successful synergies to develop growth, but many don't understand if and how these will be achieved. So, how does that affect your PMI team?

In the Engage and Examine phases, you need to get some guidance on the synergies that the buy-side company assumed to be on the cards before the target company was acquired. Then, you need to understand what, if anything, will be adapted as part of the Day One transformation, or will everything be left for Day Two or Day 100? From there, you will be able to plan and allocate resources accordingly.

Technical design

Without agreed design documents, the project simply cannot proceed. Unlike traditional projects, the high-level and low-level solution design documents in PMIs are sales documents to get buy-in at all levels across all companies involved. If you mention a sales-focused approach, though, you may find technical design people feel uncomfortable. They hear the word 'sales' and feel they are in fact being sold to. But you are looking for people to see this project, and your approach, as something different, so it's worth ruffling their feathers a little.

Make sure that your technology architects:

- Can present convincingly to the various parties
- Have a passion for technology that sells their technical design work
- Are humble in nature
- Identify and tackle the unreliable services and legacy systems first
- Lead the technology design by assessing the end user/employee journey and using this as the basis for all design
- Get some hands-on experience of the system
- Obtain sign-off for the design, as this it may be needed for evidence in the future if there are issues.

Design documents are not something that only the technical team members have to worry about; they have a major impact on the entire team, timeline, and the effort and finances required for the project. They also have an impact on how the team is perceived both on an individual level, where the employees are continually assessing whether they are in safe hands, and at a project board level, as the board will be assessing the approach and any changes to the design that will increase spend and thus require their approval.

In most traditional technology projects, once the solution has been delivered, the design documents will never be referred to again, but this is highly unlikely in a PMI. Once the company technologies have been transformed, employees will refer back to the design in order to establish what was agreed to compared to the level of service they are experiencing. The buy-side company will never be 100% certain of what technology they are inheriting until it has been operating for at least for a whole business month. Likewise, the target company cannot be sure how their technology will work in the new environment until it is implemented. Therefore, it is worth considering talking to the lead technology architects and ensuring that they stick to these guiding principles:

Keep it boring. I stole this one from a key stakeholder in a recent project, but it sums things up neatly. The technical execution/migration efforts should go unnoticed by the internal and external users of the systems, and therefore be deemed 'boring' by all those involved.

The solution is a collaboration, not a presentation. Making the design something that is presented creates a risk of passivity from at least one of the parties. There are obvious issues with a passive partner. It's easy for them to point fingers when something in the plan goes awry, but there is also the concern that a passive partner will sit on essential information that will determine the fate of your project.

The design is likely to impact technology employee engagement and buy-in to the plan, so it is essential that it is handled as a collaborative task that requires constant input from all parties. This means there needs to be clear ownership of the design with a view of cross-party responsibility, which depends on how well the parties work together.

Engage the SMEs/engineers. The joint project team must ensure the design process is driven by the right calibre of technology architects – those who can work between opposing stakeholders and gain respect from the SMEs. The project will move slowly without SMEs or engineers being involved in the design of the solution, but SMEs and engineers are likely to have differing opinions on how things currently work, and how they should work in the new world.

Use the risk register. Instead of trying to fix technology issues, consider moving them to the risk register where possible. For instance, if the team notices a business system is running slower than it was prior to the change, they can choose to spend time fixing it, or assess it and then add the perceived issue to the register as an accepted risk. This approach is often a lot cheaper.

Keep the teams small. This guiding principle is important at the technical design stage. It will prevent scope creep and maintain confidence in the technical solution because you will have fewer people to convince and can control the messaging up and down the organisational hierarchy.

Fifteen-minute meetings. This is my favourite guideline. Project managers who drive fifteen-minute 'pit-stop' meetings for the team can help prevent information overload and lethargy.

Consider the licensing costs as part of the design. If you were to engage a property architect, they would typically present the design artefacts to you along with some indicative costs. You would then assess the design's worth to you and decide whether to proceed.

In contrast, for technology projects, software licensing costs are often left out of the finalised technical design documents. The exception is if your company utilises TOAGF-certified technology architects. However, as part of the design there will certainly be some software licensing costs. If you can at least estimate these, you will have a better technical design. It makes sense to add a section to the high-level design document that provides an indication of the licensing cost implications of the project. Similar to a property improvement project, there will probably be a difference between the early estimations and the actual costs incurred, but documenting indicative costs is a good start.

Transitional service agreement strategies

Divestments may need a TSA. Clarity is required from both parties to ensure an effective TSA, so determining the five Ws of a TSA is essential:

- Why is a TSA being put in place?
- What services are to be delivered and managed?

- When will the services be provided, and critically, for how long?
- Where will services be provided?
- Who will provide services?

While it is not your job to manage the testing team, it is worth forming some agreements between yourself and the testing team manager.

Ideally, solution testing will be performed by a dedicated team. The testing team manager's ability to drive their own project will impact your success, so it is worth proactively developing a test plan that will run alongside and complement your project schedule. All too often, PMI project managers don't realise the importance of testing until a lack of it prevents a go-live, and then the whole project schedule unravels, along with employee confidence.

Be prepared for the testing team manager to be busier than yourself – they will be communicating to employees across the two businesses, testing IT components that will fail, and potentially be managing the remediation actions.

Some recommended approaches to agree on are:

- Solution testing should be performed by experienced testers (and not by the person or team that developed the solution or software, as they may miss issues)
- Ensure teams align results before reporting them back to the board
- Work from the same list of applications, specifically the corporate service catalogue

- Ensure testing is an exercise of thorough creativity, work, methodical recording, and remediation
- Develop a graphical tester map of the key people from IT and the business who will be critical to the testing process
- Testing must be performed in a testing environment

For more guidelines, please visit: http://info.beyond-ma.com/Testing

The testing environment. Your team needs to ensure that testing is not used as leverage to slow down the project delivery. Therefore, it is critical to understand and document, at a high level, what the current test environment is, as you will need sign-off from all parties before you can press the 'Go' button on the Day One transition.

Some of the basic information you need to document:

- Is a fully operational, standalone testing environment available?
- How effective is the testing environment in reality?
- Who maintains the testing environment?
- Are production and testing environments aligned?
- Is the testing environment subject to its own change control?
- How are fixes managed so as not to damage the testing environment?

If the testing environment is in fact production, the remainder of these questions can be skipped, but still note this in your testing environment report.

- Is the testing escalation/reporting process documented?
- Does everyone know when to escalate a testing issue?
- Is every core or critical service replicated in the test environment?
- Is every client desktop operating system available?
- What testing tools are being utilised?
- Are testing automation tools being used?
- What are the core or critical applications that need to be re-tested for each major change?
- Are core/critical applications re-tested every time a change is made to the environment?
- What reports will be expected from the testing team on a daily basis?

This is not a full audit of the testing capabilities of the team as that will be assessed by the testing team manager, but the above will provide a good grounding to determine risk.

In my own experience, a lack of software testing almost stopped a car manufacturing line. The entire joint team had gone through the motions of software development and testing, but the employees hadn't been engaged.

While testing is a major concern for the company and acquisition, it is also a potential threat to your own career and reputation.

So, how can you avoid this threat? By focusing on the staff, not yourself. Testing is overlooked in many ways, so make sure the staff have the resources to do their job. Engage the testing staff and help them find opportunities to learn during the project. Get buy-in through engagement, planning and design with adequate room to cater for known M&A integration issues.

Integrated testing. If an entire technology environment is being transformed or absorbed, it is likely that multiple software changes will slip in with some of the technology platforms. This can represent an impossible mission for the testing team as it's a moving goal post – where do they start? What and whom do they focus on? They are likely to be so busy putting together the testing plan that they will need someone else to look at the bigger picture.

Some testing guidelines:

- Prioritise applications, highlighting the critical applications that require sign-off
- Do not change the underlying technology platforms unless absolutely necessary
- Phase changes in earlier than Day One if possible
- Make use of automated testing tools and script
- Dedicate staff to services, application and desktop testing

Testing in different locations. Ensure the testing team is testing the Day One solution in all office locations in which the services will be utilised, otherwise there is a high chance of the solution appearing to be a failure at satellite sites due to response times.

Testing in different locations does not necessarily mean taking the time to travel to different offices physically: you can use remote technology and ask people at each location to perform a set of pre-defined actions while your team facilitates or monitors the outcome.

Dress rehearsals

A dress rehearsal is a practice of the Day One integration. On the surface, a dress rehearsal allows the team to ensure the Day One integration plan and technology solution will operate as expected. It differs to IT testing streams as it is run by the core cutover team rather than the testing team, mostly because the cutover team will simply apply the cutover solution rather than engineer it on the fly. This way there are no sticking plaster changes being made during the test runs, and any issues are highlighted and reported.

A dress rehearsal is an extensive piece of work that utilises already stretched resources. The cutover team will likely be headed by members of the infrastructure team, but the dress rehearsal will also require the involvement of a small number of key people from each party, along with major applications.

Why perform a dress rehearsal? A dress rehearsal is a fantastic way to force the team to work together to find issues. It is essential to give the board confidence in the team and their approach. The main aim of a dress rehearsal is to convince the internal team and the project board that you will be able to deliver the technical solution within the time slot permitted. From a technical perspective, the objective is to eliminate assumptions around the process timings and capabilities.

Assuming you have a testing environment that you can use for the dress rehearsal that fully represents (or almost matches) the production environment, there will be some additional benefits:

- The dress rehearsal will allow you to demonstrate how the environment will perform and operate in the longer term
- You will create a more realistic testing environment for the testing team to use
- It provides the ability to run some data and application integrity checks – for example, will databases or customer information be correct after the transformation?

To gain buy-in across the group and clarity of approach, the team needs to develop and publish a dress rehearsal approach and design document. Not only do you need to deliver and assess a test cutover plan, you must also remember to test a fall-back plan which will be used if there is a failure. This must include service availability, security and data integrity checks.

When to plan a dress rehearsal. The best time to start planning your dress rehearsal is soon after the high- and low-level technical design documents have been signed off, as this will provide plenty of time to go into the necessary detail and get input from all the stakeholders across the teams.

Who needs to be part of the dress rehearsal? Using a small team makes it easier to keep track of progress. If this small team is provided with some guiding principles regarding how to approach the dress rehearsal, they can keep track of and report on any potentially risky amendments made or planned during the process.

It is tempting to include the majority of the technology delivery team in the dress rehearsal, but for it to be more effective, create a small sub-team that consists of members from the project management, software development, infrastructure, and service management teams.

How do you carry out a dress rehearsal? If it's phased, this is potentially a luxury, as you can migrate sections of a business IT infrastructure and treat it as a BAU exercise.

It is not uncommon to work with a client who has decided to complete a PMI project many years after the M&A deal was completed, usually because the acquired business was loosely-integrated with the buy-side and operating sufficiently independently. PMI projects then become more like technology consolidation efforts looking to reduce costs by bringing systems in-house. This is especially common due to private cloud systems and 'infrastructure as a service' platforms like Amazon Web Services and Microsoft Azure. I am sure phased PMI work will continue.

Phased PMI may be run by BAU staff without the traditional project control documents. This is fine until things go wrong. It's difficult to track issues without someone overseeing the process and documenting RAID logs. It may also mean that the PMI activity is occurring underneath the radar, which can create difficulty engaging stakeholders, negotiating change acceptance and obtaining budget as there's no urgency or sponsor.

The key difference with a big bang approach is the level of detailed planning you will need for both the dress rehearsals and the actual cutover itself. Tasks and actions will need to be examined on an hour-by-hour basis, and you'll need to test your different streams of migration and ensure certain milestones are being hit. Therefore, planning is the most essential tool, and once again, careful work stream planning is needed.

If the cutover is a big bang exercise for a specific date, it adds a lot of pressure to the process, the approach and the plan. Firstly, everything must work on Day One, so you must get that across to your team. Secondly, the business and its customers have to know that things will be out of action for a while. You need to communicate and put measures in place to ensure people are fully informed of the situation and understand when the services will resume.

Regardless of which of the two approaches you take, you must ensure that no application updates or system upgrades are planned for the time of a dress rehearsal unless they're key to the project. If you are making any application or data changes during the intended cutover, there is a risk that the result will be very different to what was shown during your testing. In extreme cases, this may invalidate any of the testing you've done prior to the cutover, meaning you'll need to re-execute the whole project.

It is not unusual to run three to four dress rehearsals prior to cutover, and to end up with slightly different results every time. I would encourage you to expect at least one or two dress rehearsals before the actual migration. This will allow you to work out where the gaps are between the teams, and you'll identify whether there are going to be any issues with supporting services or applications.

If you have more than two dress rehearsals, you will eliminate some of the risk. But there's a chance results will be different each time you run the dress rehearsal, so you may need to run even more. I would suggest telling the board that more dress rehearsal activities will happen than you've planned for.

You will need to consider how you are going to communicate in a fast-paced dress rehearsal process. If you keep the team small, this is simple. Real-time communication is preferable as staff will often be in different locations and data centres, so something that is immediate, wide-reaching and secure is suitable. I have seen companies use WhatsApp in encrypted mode, and more recently Slack and Microsoft Teams. But your choice of tools may be restricted by corporate security requirements from all parties. You also have to make sure that any type of communication is encrypted and secure if sensitive company IT information will be transmitted.

The key is to communicate in a way that is agile and swift and doesn't take up too much of the team's time, but enables them to keep each other informed. Consider the details long before your dress rehearsal so the team will not have to think about basic communication tools on the day.

Summary

The Envision phase is where you build the most significant assets of your PMI project – namely, the technical designs and project plans. Without either, the project cannot proceed.

The art of designing and planning is to spend the right amount of time on it and drive the team to deliver the most appropriate, and perhaps even innovative, solutions and approaches. But plans are nothing without an element of testing, research and development, and practising the Day One integration. Without a hands-on approach, everything is simply theoretical.

With all the components in place, you can commence to executing the Day One transition.

Conclusion

The main thing is to keep the main thing the main thing.

STEPHEN COVEY

Great change is seldom done in isolation; it usually takes a team of dedicated people, and so often in corporate technology projects, those people are represented as rectangular blocks on a Gantt chart or as a list of names in a spreadsheet. They are simply things project managers move about as resources.

One definition of resources is:

> *A stock or supply of money, materials, staff, and other assets that can be drawn on by a person or organisation in order to function effectively.*
>
> Wikipedia

You can see the order of importance of resources – money, inanimate objects and *then* staff. A 'people last' approach to project management seems to be the general attitude when the pressure of a major project suddenly hits those responsible.

When was the last time you heard a project team say that their abilities, personalities and needs were the foundation for the change? I am guessing never.

In the technology industry, we tend to dissect, dismantle and examine software, systems and data intimately. We sniff network packets, debug lines of code and spend fortunes ensuring the user experience is just right. We hardly ever make the assumption that throwing a load of code together will work, and we put people and processes in place to check the quality throughout. But in contrast, when it comes to people – who individually are a gazillion times more complicated than any computer system – we treat them as black boxes.

> *A black box is anything having a complex function that can be observed but whose inner workings are mysterious or unknown.*
>
> Collins English Dictionary

Imagine technology employees of the past walking, interacting, working together as self-contained units. These are the black box people, and we generally never examine them, certainly not to the degree we examine technology, until something goes wrong. When two or more black boxes stop working, we're in trouble, and if we have no understanding of their inner workings, how on earth can we fix them?

If we spent a fraction of the money already spent on quality assurance software or systems for monitoring, on analysing and monitoring our own teams, we would already be far further ahead than most. The two main ways of assessing staff psychometrically cost less than $100 each, which is nothing compared to the cost of software. As there are so many M&A tech project failures, why not spend a little bit of money and time to do something differently to get different results?

I am suggesting you assess people before things go wrong. And people will probably be resistant because they've never been tested in this way before. It's not common practice to do this as part of a technology projects.

It is worth noting that the digital natives, i.e. Generation Y and Z, are extremely data driven, so will not be so averse to taking these tests. I have met start-up entrepreneurs straight out of college who have already implemented psychological profiling for their entire team. While you may not currently be taking this approach, you may need to in the future to compete in your field.

If you have followed the steps outlined in this book, you will already have assessed your team members as individuals. Instead of making your first written words regarding your PMI project about computer systems, business cases or project initiation documents, you will have written about your team members as people with their own conative profile and technology skills. Not only will you have found out more about the people, you will understand their differences. You can then utilise this information to form a better, more dynamic team to tackle the journey ahead.

There is also benefit in demonstrating to your staff that you care about them, their current roles and how they may perform during a tough and maybe adverse transformation project. And by showing that you care, you may just limit the loss of all the good people in the team.

What happens next?

Simply put, PMI projects tend to fail because of lack of interest from the technology employees, poor identification of systems, and terrible plans. Throughout the Engage, Examine and Envision phases, you can develop better relationships and engagement from your joint technology team, then move on to delivering a far more effective technology due diligence to identify and highlight roadblocks and opportunities early. Lastly, you can develop a project plan and technical design that will sell your PMI transformation.

It sounds corny, but this is genuinely only the beginning. The next stages are to deliver your plan to reach Day One, and once that's over, plan for the evolution of the companies and teams involved from Day 100 and beyond.

To find out more about the next steps, the 'Evolve' and 'Execute' phases of the Managed M&A, take a look at http://info.beyond-ma.com/bonus-chapters

Trust is key

I was visiting the famous Taj Mahal in India with a friend. The queues were enormous and the idea of having to wait in the blistering heat without cover was unbearable. Even though I appear Indian in appearance, it was obvious that I was a tourist, and someone offered to move us up the queue. For a fee.

We paid the fee, and suddenly we were minutes away from entry. But then someone else approached me and explained that they didn't allow mobile phones and a plethora of other items inside. For a fee, he would take the items off me, and once I'd finished my tour, I would need to look for him and he would return the items through a hole in the wall.

I looked the man in eyes. In those nanoseconds, I trusted him enough to hand over my phone and other paraphernalia. My phone at the time would have been worth a year's wages for him.

I was enthralled by the spectacular view, the feel, the history of the Taj Mahal. But two thirds of the way round, I had a sudden pang of worry. Why did I hand over my phone?

As I left the Taj, there was the man waiting patiently for me. He gave me a smile and handed me my items through the hole as he'd promised. The fee I'd given him was peanuts compared to what I had trusted him to look after. But something had told me it'd be alright.

Whatever you do, trust your instincts. And trust people.

Glossary

Acquisition	The business being acquired, or the process of taking over another company.
API	Application Program Interface; in the context of PMI, APIs may be used to allow one program or service to talk to another in a different business.
Carve-out	Another term for divestment; where an entire business unit is 'carved out' of its origin for integration into the purchaser's business environment.
Continuous improvement	An ongoing effort to improve products, services, processes, or skills.
CMDB	Configuration management database; the database of a firm's technology assets.
Cutover	A rapid transition from one phase of a business enterprise or project to another.
Day 100	The first 100 days of a merger have a very high impact on the overall success of an integration.
Day One	In the context of PMI, the need for the business to be ready on Day One to deliver services and products to customers, with no impact on current customer value and cash flows.
Divestment	The action or process of selling off subsidiary business interests or investments.
M&A	Mergers and acquisitions; many see all activity in this market as acquisitions because even with a merger there is usually a dominant party in the deal.
Middle-ware	Software that customers never see but is essential to allow communication and transactions.
Novation	Moving contracts from one party to another; especially important for existing software contracts.

PMI	Post-merger integration.
R&D	Research and development.
RACI matrix	Responsible, accountable, consulted, and informed: a checklist used to clarify roles and responsibilities in cross-functional/departmental projects and processes.
RAID log	A record of risks, assumptions, issues and dependencies; an essential tool for managing risk.
Reverse takeover	A takeover of a public company by a smaller company. In PMI, it may be possible to consider this approach when the staff of a digital acquisition 'takes over' the reigns of tech in a larger company.
Risk register	A log to manage risks; unlike a RAID log, which is project-oriented, a risk register may cover the entirety of an organisation's technology environment.
Strategic fit	How well the mission, products and services of an acquisition meet those of the purchasing business.
Synergy	The assumed (potential) financial benefit achieved through the combining of companies.
TUPE	Transfer of Undertakings (Protection of Employment). The legislation that protects the terms and conditions of employees moving from one company to another.

Acknowledgements

Every young person needs a little bit of luck and confidence to succeed, and three people really helped me immensely:

- None of this would ever have happened if my brother Trevor hadn't been so generous in giving me my first computer, a Sinclair ZX-81, back in the late 70s. You helped me develop a 'bad habit' and have been supportive ever since.
- Dentist Shah rebuilt my teeth and mouth over three months of (agonising) surgery, and then offered to pay for the entire treatment – what an amazing gesture of kindness.
- Then there was Reg Thompson, IT director of thousands of people at Ford, who wanted to make a point by giving me, an uneducated non-graduate, a graduate job. You took a risk on me that transformed my life.

Once computers, confidence and a career were all lined up, my life changed significantly.

Thank you to my beautiful wife Natasha – you have supported me throughout my challenges as I set up a business. Your creativity inspires me and everyone around you daily.

Lucille and Juanita: you've had to endure listening to years of me on long-distance calls as I've developed my real 'people skills'.

Thanks also to Kathy Kolbe and her amazing team at Kolbe Corp. in Arizona. Kolbe has helped our team and our clients in so many ways, and we are excited to have them fully bedded into our business.

Thanks also to Sue Clayton, the filmmaker who allowed me onto her screenwriting Masters' programme – certainly two of my best years in London.

To Daniel Priestley and his team at Dent Global, whose accelerators saved me from taking unnecessary qualifications and, instead, made me look inside to work out what I already had to offer. I love KPI!

To Lucy McCarraher and Joe Gregory and their team at Rethink Press, who demonstrated how 'easy' (!) it was to develop a book and bring it to life.

To everyone at our team at Beyond – what a party! Special thanks to Dave Refault for (a) putting up with my 'crazy' ideas and (b) providing blunt feedback on this book – it worked.

And to Misty Henry – I hope you also get the opportunity to travel the world and take a 'people first' approach to everything you do in life.

The Author

Hutton Henry is currently the Managing Director of Beyond M&A, a technology consultancy based in London and specialising in psychometric assessments and post-merger integration. He has worked on many merger and acquisition integration projects for companies including HP, Ford, and Jaguar. In 2012, Hutton's first consultancy was recognised by Microsoft as a Global Top Three Partner for Cloud Technology within Enterprise Business. He has also trained with Nancy Kline, the founder of Time to Think and the Thinking Environment, and Kathy Kolbe the psychometric expert.

Hutton has a deep belief that creativity, business, and technology literacy provide people with a fantastic ability to grow and make themselves more secure – but only if people become creators of technology, not consumers. All the proceeds from this book will therefore go to CodeClub, a charity that helps 9–13 year olds to learn software development skills, no matter what their background.

Currently, Hutton's company is working with customers to deliver 'people first'-oriented merger technology projects, all based on his thirty years of industry experience.

Contact Hutton at:
Hutton@Beyond-MA.com
www.beyond-ma.com